The Plants of Dom, Bamenda Highlands, Cameroon

— A Conservation Checklist —

Martin Cheek, Yvette Harvey & Jean-Michel Onana

Royal Botanic Gardens, Kew
IRAD-National Herbarium of Cameroon

Kew Publishing
Royal Botanic Gardens, Kew

PLANTS PEOPLE
POSSIBILITIES

First published in 2010 by
Royal Botanic Gardens, Kew,
Richmond, Surrey, TW9 3AB, UK
www.kew.org

ISBN 978-1-84246-398-7

British Library Cataloguing in Publication Data
A catalogue record for this book is available from the British Library

Typeset by Yvette Harvey (Main Checklist and Introductory Chapters)
and Janis Shillito (Introductory Chapters)

Cover design by Christine Beard
Publishing, Design & Photography
Royal Botanic Gardens, Kew

Front cover: *Clerodendrum carnosulum.* Photo taken by Yvette Harvey, February 2006.
Back cover: Dom, Cameroon, decrease in forest extent, by Susana Baena.

Printed in the UK by Hobbs the Printers

For information or to purchase all Kew titles please visit
www.kewbooks.com or email publishing@kew.org

Kew's mission is to inspire and deliver science-based plant conservation worldwide, enhancing the quality of life.

CONTENTS

LIST OF FIGURES

ACKNOWLEDGEMENTS

The following abbreviations are to be found in the text: ANCO (Apiculture and Nature Conservation Organisation (formerly North West Bee-Farmers' Association – NOWEBA)); BHFP (Bamenda Highlands Forest Project); DED (German Development Organisation; FHI (Forest Management Institution; GIS (Geographic Information System); HELVETAS (Swiss Development Organisation, Cameroon; HNC (IRAD-National Herbarium of Cameroon); HSBC (HSBC Bank Plc); IUCN (International Union for Conservation of Nature and Natural Resources); IUCN-NL (IUCN National Committee of The Netherlands); K (Herbarium, Royal Botanic Gardens, Kew); Kew (Royal Botanic Gardens, Kew); KIFP (Kilum-Ijim Forest Project); LBG (Limbe Botanic Garden); MINEF (Ministry of the Environment and Forests); NEPA (Natural Environment Protection Agency); SIRDEP (Society for Initiatives on Rural Development and Environmental Protection); SNV (Netherlands Development Organisation, Bamenda, Cameroon).

Firstly we wish to thank those who facilitated our fieldwork at Dom Community Forest: the Fon of Dom and his elders and chiefs are thanked for receiving delegations from our expeditions, and for showing interest in the progress of our work; the Divisional officer in Nkor, an advocate of forest conservation, who very keenly supported our work.

The Apiculture and Nature Conservation Organisation (ANCO) (formerly North-West Beekeeping Assocation – NOWEBA) at Bamenda, particulary Paul Mzeka, who "commissioned" the botanical survey of the Dom community forest, suggested using Dom village as a base, introduced us to its community and provided logistical support and ANCO staff and their participation in the field. Paul Mzeka, director of ANCO heads WHINCONET (Western Highlands Nature Conservation Network), a consortium of NGOs whose objectives are biodiversity conservation and poverty alleviation.

Teams comprised:

2005: ANCO: Walters Cheso & Kenneth Tah; HNC: Mme Nana; Kew: Marcella Corcoran; Other participants: Verina Ingram (SNV, Netherlands Development Organisation, Bamenda, Cameroon), Benedict J. Pollard; Community Forest Guards, Tantoh Sale & Victor Nfon

2006 (February): ANCO: Kenneth Tah & George Kangong; HNC: Mme Nana & Dr Onana; Kew: Marcella Corcoran & Yvette Harvey; Community Forest Guards, She Frederick, Kasimwo Wirsiy & Johnson Nchianda; Quinta Bousaw (cook); Abraham Njume (driver, Presbyterian General Hospital, Nyasoso)

2006 (September): ANCO: Kenneth Tah; HNC: Victor Nana & Olivier Sene; Kew: Martin Cheek; Fon of Dom, Eric Chia, Community Forest Guards, Rene Ndifon, Kasimwo Wirsiy & Johnson Nchianda; Jude Zenebiun (driver, ANCO); Eric Chia

At the National Herbarium of Cameroon, Yaoundé, we thank the present leader, Jean-Michel Onana, and the following, whom have joined us in fieldwork Dom and/or identified specimens: Barthélemy Tchiengué (Researcher);Victor Nana (Technician); Felicité Nana (Accountant); Olivier Sene (Technician); Blanche Nke (Technician).

We would like to thank Julie Mafanny of LBG who facilitated Elias Ndive's visit to Kew in Spring of 2008. Elias, amongst a myriad of other work, helped identify expedition material.

Our fieldwork is not possible without the generous ongoing sponsorship and assistance from the Darwin Initiative (September 2006 expedition), the Bentham-Moxon Trust (Marcella Corcoran's 2005 & 2006 expeditions) and a Kew Overseas Fieldwork Committee grant in 2006 (February).

The Darwin Initiative grant to Kew has been the main single source of funds to our 'Red List Plants of Cameroon' Project (http://darwin.defra.gov.uk/project/15034/), from which this volume is one of the outputs. We are extremely grateful for this assistance, without which the fieldwork, specimen identifications and species descriptions which form the basis of this book would not have been completed. A substantial part of the publication costs of this book were met by this grant. In particular we thank Ruth Palmer and Helen Beech for their administration of this grant at DEFRA, as well as George Sarkis, financial accountant at Kew for assistance in managing the finances at Kew. The Edinburgh Centre for Tropical Forestry, particularly Eilidh Young, contracted by DEFRA to monitor the grant, are thanked for their constructive reviews of our project throughout its life.

At Kew we thank the unstoppable Marcella Corcoran who joined in the fieldwork, and played a leading role in training staff in horticultural techniques at the Kitiwum nursery. Of all the Kew staff, Marcella has spent the most time exploring Dom, visiting in 2005 and 2006. In spite of cracking a rib during a terrible fall on the first day of exploring in 2006, Marcella continued collecting, in good humour, with the group for the duration of the trip.

The Publications Committee agreed to contribute towards the publication costs and arranged review of the manuscript. Lydia White and John Harris are thanked for guidance on producing the Camera-ready copy from which this book was produced, and for arranging scanning of images and design work.

At Kew we also thank David Mabberley, Keeper of the Herbarium, Library, Art and Archives (HLAA) and Simon Owens, former Keeper of the Herbarium, who has long supported our work in Cameroon, as has Rogier de Kok, Assistant Keeper for Regional teams, who has also championed our work. Eimear Nic Lughadha has also been consistently supportive of our efforts in Cameroon. We also thank Daniela Zappi and her support and guidance in her role as the previous Assistant Keeper under whom this project was run. Laura Pearce and Xander van der Burgt have provided valuable sectional support within our team, facilitating the time that the Kew authors have been able to devote to this project. Also to be thanked are Susana Baena for producing the maps for the back cover, and Chris Beard who designed the overall cover and plates.

Determinations of the specimens gathered in the course of our fieldwork were made by ourselves and by the following botanists, often world experts in their fields. We sincerely thank them all (in checklist order), and their institutions, for their work, often done without reservation, towards producing the names which are used in this book.

We thank (in checklist order) B. Pollard (Araliaceae & Labiatae), M. Thulin (Campanulaceae), O. Sene (Capparaceae, Euphorbiaceae & Moraceae), E. Ndive (Crassulaceae, Euphorbiaceae & Moraceae), B. Nke (Crassulaceae, Euphorbiaceae & Passifloraceae), B. Tchiengué (Euphorbiaceae & Monimiaceae), S. Williams (Euphorbiaceae), M. Vorontsova (Solanaceae), P. Boyce (Araceae) and A. Muasya (Cyperaceae).

At Kew we thank (also in checklist order) I. Darbyshire (Acanthaceae), G. Gosline (Annonaceae), D. Goyder (Asclepiadaceae), H. Beentje (Compositae), N. Hind (Compositae), L. Pearce (Cucurbitaceae, Menispermaceae, Passifloraceae & Vitaceae), B. Oben (Euphorbiaceae), A. Paton (Labiatae), G. Bramley (Labiatae), B. Mackinder (Leguminosae: Caesalpinioideae; Mimosoideae; Papilionoideae), B. Schrire (Leguminosae: Papilionoideae), R. Polhill (Loranthaceae), T. Utteridge (Myrsinaceae), E. Fenton (Passifloraceae), S. Dawson (Rubiaceae), M. Wilmot-Dear (Urticaceae), A. Haigh (Araceae), K. Hoenselaar (Cyperaceae), P. Wilkin (Dioscoreaceae), T. Cope (Gramineae), D. Roberts (Orchidaceae), P. Cribb (Orchidaceae), W. Baker (Palmae) and P. Edwards (Pteridophyta).

A special mention has to go to Dave Roberts, one of Kew's orchid specialists. With only a couple of days before going to publishing, Dave helped confirm determinations for orchids from images of specimens retained at HNC. Unfortunately orchids are proving more problematic for us to handle within the checklists. Ironically, the need for CITES export and import permits, means that a crucial group of plants for a conservation checklist have nearly missed being identified and thus included in the work. Dave Roberts (see http://www.kew.org/kewscientist/ks_33.pdf & http://www.cites.org/common/com/PC/17/X-PC17-Inf-06.pdf, both accessed 30.vi.2009) has shown that it is because of the problems encountered exporting and importing material needing CITES permits between institutions that are not registered, that fewe orchids are being collected. Regrettably HNC is not a CITES-registered institution, and so export and import permits have to be acquired (the complicated process is explained after *Bulbophyllum calvum* in the main checklist). Recent changes in DEFRA also means that importing orchid specimens has become prohibitively expensive for projects. Ignoring the expense of the export permit from Cameroon, to pay for the import to the UK, the orchids collected for the checklist (7 genera) will cost £413 (£59 per genus) (http://www.defra.gov.uk/animalhealth/Charges/citesfees.htm, accessed 30.iv.2009) to import.

Gratitude is to be given to The Andrew J. Mellon foundation for its continued support to http://www.aluka.org/. Type material stored around the world is instantly accessible as is important literature, speeding up identifications and finding references. In fact, scanning and computing equipment, purchased with a grant from the foundation has proven very useful for naming the orchids (see above) and other unicates stored at HNC, along with exchange of data and accounts for editing.

A huge thanks goes to Stuart Cable of Kew. For the Mount Cameroon Project in 1997 and 1998, he originated the species database from which the checklist part of this book was produced. George Gosline subsequently developed the 'western Cameroon specimen' database, enabling us to print off data labels whilst still in the field, a real time-saver. He also developed the system by which data could be exported from the Access database using XML and XSLT into its Microsoft Word format. For meticulous data entry, Karen Sidwell, Suzanne White, Julian Stratton, Benedict

Pollard, Emma Fenton and Harry de Voil are also to be thanked. Without the database this volume would have taken considerably longer.

Craig Hilton-Taylor, IUCN officer for Red Data, based at WCMC in Cambridge, is thanked for reviewing the Red Data assessments.

An enormous debt of gratitude goes to Janis Shillito, one of Kew's volunteers, for typing Martin Cheek's Introductory Chapters. Harry de Voil spent a few weeks making editorial changes to the taxon accounts stored on the database and is to be thanked for speeding up the later editing process.

NEW NAMES

The following are published in this volume for the first time:

MYRSINACEAE

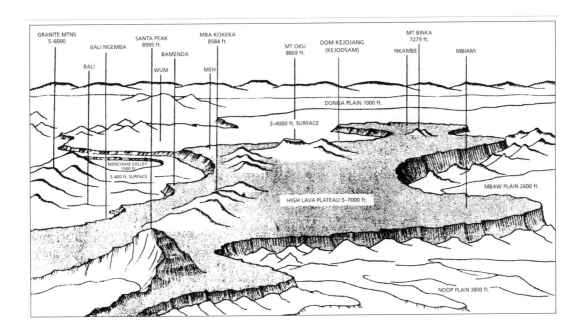

Fig. 1. The Bamenda Highlands area. View from the south with Dom arrowed. Reproduced and modified with permission from Hawkins & Brunt (1965).

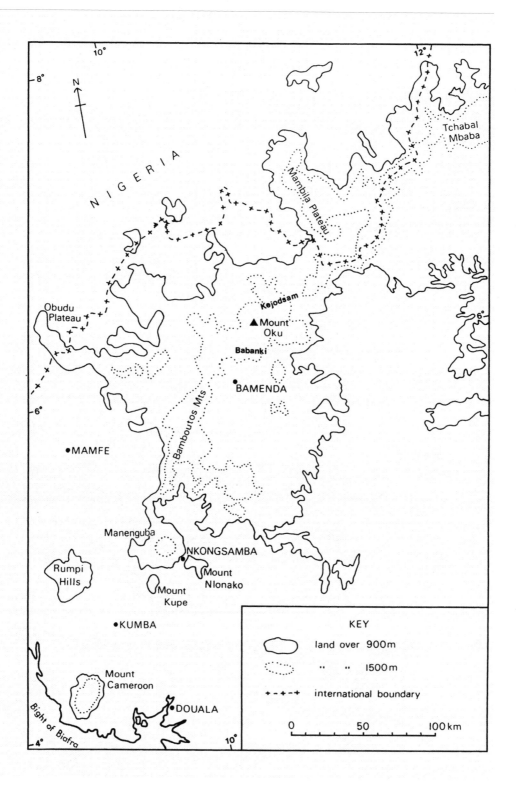

Fig. 2. The Cameroon Highlands, showing 'Kejodsam' (the Kejojang forest area including Dom).
Reproduced with permission from McLeod (1987).

FOREWORD

Martin Cheek
Herbarium, Royal Botanic Gardens, Kew, Richmond, Surrey, TW9 3AE, UK

Dom is located in Noni subdivision, NW of Kumbo in the Bamenda Highlands of North West Region (formerly North West Province) of Cameroon (Figs. 1-3). The Bamenda (or Western) Highlands themselves form part of the Cameroon Highlands which run from Bioko and Mt Cameroon in the south, northwards to Tchabal Mbabo and then continue eastwards as the Adamaoua Highlands, following a geological fault line.

It has been estimated that 96.5% of the original forest cover of the Bamenda Highlands above 1500m altitude has been lost (Cheek *et al.* 2000: 49).

Few have heard of Dom, even within the Bamenda Highlands themselves. There are those who will ask 'Why is this village important enough to merit a book?'. The answer is that Dom is home not just to several hundred people, but to several areas of forest now protected, managed, and in the process of being enlarged by the people of Dom, guided by the Bamenda-based NGO Apiculture and Nature Conservation organisation (ANCO). The story of ANCO's link with Dom and the process of forest restoration at Dom is documented in several of the following chapters in this book by Paul Mzeka and Kenneth Tah.

These 452 hectares reserved for forest and forest restoration at Dom are among the last fragments of a much greater area of forest that occurred in the Kejojang Mts (Kejodsam of maps see Fig. 3). Satellite image analysis has shown that between 1988 and 2003 this once great area of forest was reduced in area by c. 50% (see rear cover article by Susana Baena) from c. 2274ha to 1242ha. If this clearance of forest for agricultural purposes continues, in a few years, none of the Kejojang forest will remain apart from that at Dom and problems with dry-season water supplies for local communities could worsen further. Moreover, as this book reveals for the first time, numerous rare species of plant will be lost to mankind forever if this forest disappears.

This book describes the 356 species and varieties of plant discovered in the forests of Dom. These were uncovered during three short surveys at different seasons in 2005-2006 by teams of British and Cameroonian research botanists, when 376 specimens and over a hundred sight and photographic records were made. Following study of these at the Royal Botanic Gardens, Kew in London by international specialists, 23 globally threatened species were found to be present. These are discussed in detail in the chapter on Red Data Taxa. Dom is home to the world's largest and therefore most important populations of the critically endangered trees *Newtonia camerunensis* and *Oxyanthus okuensis*. Several of its Red Data taxa, such as the tree *Antidesma pachybotryum* do not occur in any formal protected area anywhere in the world.

Twelve species new to science have been described or are due to be described shortly from material discovered at Dom (see Endemic, Near Endemic and New Taxa). At least five of these new species appear to be unique to Dom and are known nowhere else in the world, one of which, *Ardisia dom*, is described for the first time in this book. Another, currently in press elsewhere, is Africa's only uniquely tree-dwelling sedge, *Coleochloa domensis*.

Elsewhere in the Bamenda Highlands, forest is secure at Kilum-Ijim (Mt Oku and the Ijim Ridge), but that is montane forest, between 2000m and 3000m altitude. The more species-rich submontane or cloud forest (c. 800-2000m altitude) of the Bamenda Highlands, once the dominant vegetation, is all but totally replaced now by grasslands and *Eucalyptus* plantations. Only one fragment, Bali Ngemba Forest Reserve, south of Bamenda has any official protected status and this has been at low level only (Harvey *et al.* 2004). As shown in the chapter on Vegetation, there are dramatic differences between the species composition of the submontane forest at Dom and that at Bali Ngemba: underlining further the need to support protection of the forest at Dom.

This book is the fifth in a series of 'plant conservation checklists'. Its predecessors dealt with the plants of Mt Cameroon (Cable & Cheek 1998), of Mt Oku and the Ijim Ridge (Cheek *et al.* 2000), Kupe, Mwanenguba and the Bakossi Mts (Cheek *et al.* 2004) and of Bali Ngemba Forest Reserve (Harvey *et al.* 2004).

A major purpose of this book is to enable identification of the plant species within the checklist area, in particular those threatened with extinction – the highest priorities for conservation. To this end, details to aid the monitoring and management of each of the threatened taxa are given in a separate Red Data chapter, which includes illustrations for as many of the species involved as could be obtained. It is hoped that this information will aid the long-term survival of these threatened taxa.

Although the work presented here on the plants of Dom-Kejojang is the most detailed and complete account available, it is not exhaustive. It is certain that yet more important plant species await discovery at Dom. News has just arrived (July 2009) that Kenneth Tah has discovered in the Dom forests the Critically Endangered *Oncoba* sp. nov. of Mt Oku.

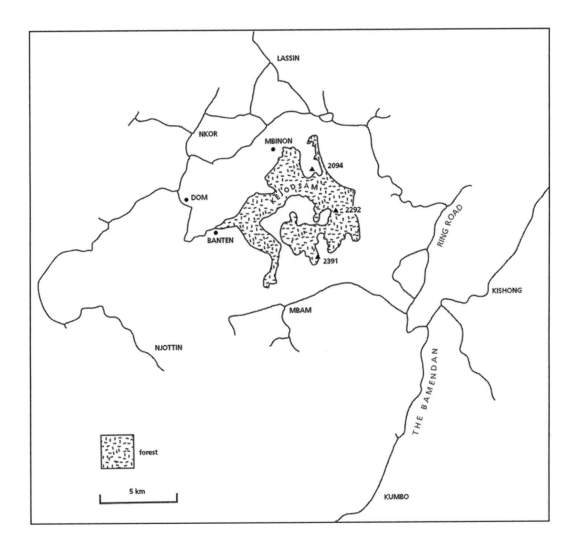

Fig. 3. The location of forest at Dom and 'Kejodsam' (Kejojang) – based on the IGN 1: 200,000 map, Nkambe Sheet – from aerial coverage 1963-1964. Drawn by Martin Cheek.

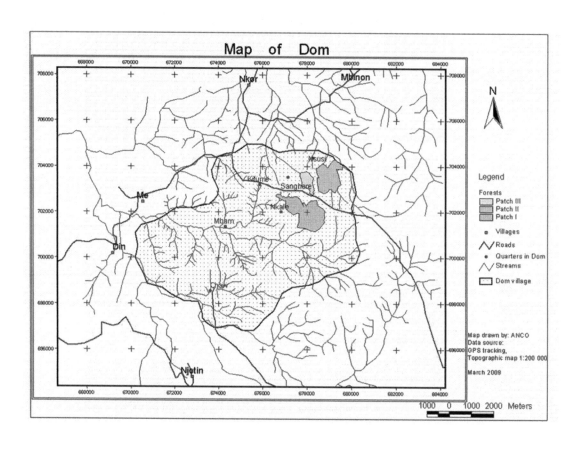

Fig. 4. Map of Dom showing its six Quarters and the three forest patches. Drawn by Kenneth Tah, ANCO, March 2009.

THE CHECKLIST AREA OF DOM

Martin Cheek

Herbarium, Royal Botanic Gardens, Kew, Richmond, Surrey, TW9 3AE, UK

The checklist area consists of three community forest patches (and interstitial areas) which lie within the Dom area as demarcated in Fig. 4.

Dom is reached from Kumbo, via Mbinon and Nkor. Nkor is the principal settlement, base of the Divisional Officer and headquarters of the Noni subdivision within which Dom lies. There are six settlements within Dom, each separated by one to several kilometres: Mbam, Nkale, Sanghere, Chaw, Nsusi and Kifume. The last of these (6°21'33"N, 10°35'34"E) was our base for exploring Dom and houses the residence of the Fon of Dom.

The forest patches are: Kowi (patch I); Nvui (patch II) and Nsusi (patch III). Dom meaning "push" was formerly known as Bvugoi. The area of these patches is 452.7ha. The range of altitudes covered by our inventories was c. 1550-1930m alt.

Kejojang (Kejodsam of the 1:200,000 Nkambe sheet and so of BirdLife reports) is the name of the former forest block, now much fragmented, that lies east of Dom and of which the former forest block, and of which the Dom forest patches form the western extremity. Further forest patches lie north of Dom and fall under the village of Mbinon. Kejojang means "gives rain" in Noni. The Kejojang forest area then, extends east of Mbinon and Dom towards the Nso tribal area. In 2003 the forest area was c. 1242ha (Baena, rear cover of this book). Dom holds about a quarter to a third of the total (2003 estimated) Kejojang forest area and appears to be the only protected part of the whole.

Note: Kejodsam is the name of the Mountain range.

A reduced version of this article was originally published in the FAO Africa magazine, Nature & Faune 23(2): 58 (Feb. 2009)

DOM, A TINY VILLAGE IN THE BAMENDA HIGHLANDS OF CAMEROON SWINGS INTO PROMINENCE

Paul N. Mzeka

Coordinator ANCO, P.O Box 5150, Bamenda, Cameroon.
Tel: +237 77 71 56 5 / 99145918
Email: mzeka.anco@yahoo.com   Website: www.anco-cameroon.org

Summary. Defined by World Wildlife Fund as one of the 200 world wide ecoregions, the Western Highlands of Cameroon stretch from the Atlantic archipelago in the Gulf of Guinea to the heart lands along Cameroon's western border region. The Highlands are rich in endemic species, particularly in the Bamenda Highlands. Unfortunately, most of the wildlife has been lost. However, the remaining forest patches still contain remnants of the endemic species of both flora and fauna. One of these is the bio rich patch at Dom, a mountain village in the northern flanks of the Kilum Mountain.

Early 2008, after completing a project to save a similar rich, forest patch in the village of Mbiame, the Apiculture and Nature Conservation Organisation (ANCO) a conservation NGO founded in 1992 and her partner NGOs launched another to save the Dom patch (452.7Ha). The objective is to relieve the intense pressure on the fragile submontane ecosystem arising from human encroachment. Also, bush fires, particularly during the dry spell, damage the fragile ecosystem.

The ongoing project has 4 components: biodiversity conservation, income generation, sustainable land management and school environmental studies. Ten sensitisation sessions, spread out at different levels for a duration of 3 months, outside the project schedule were used to thoroughly sensitise the community members about the possible project activities, their outcome and the role of the community in the project implementation.

On the basis of rich experience in community based conservation, it is argued that five principal pillars are essential for success in this field, namely: significant community involvement, good governance, income generation, community-based leadership and equitable gender representation.

BACKGROUND

The World Wildlife Fund defines the Western Cameroon Highlands range that stretches from the Atlantic Ocean archipelago of São Tomé, Príncipe and Bioko (Equatorial Guinea) and continues on the mainland through Mount Cameroon (4095m), the Rumpi Hills, the Bakossi Mountains, Mount Nlonako, the Kupe Mwanenguba, the Bamboutos, the Bamenda Highlands and beyond as an ecoregion (Ingram, V. *et al*, (2007). Networks and Networking in the Cameroon (Western) Highlands). The flora of this region "constitute one of the most unique and threatened ecosystems in Africa (and) are home to a number of endemic species of bird, amphibian, reptile, mammal and insect" (Bamenda Highlands Forest Project: The Community Based Conservation in the Bamenda Highlands, Cameroon. Annual Report 2002 and 2003). However, the ecosystems of the region particularly those of the Bamenda Highlands are under intense threat from human activities. This has caused the fast disappearance of forests and as they disappear, the benefits and services they provide also disappear. The reduced forest cover in the Bamenda Highlands has left behind isolated fragments of forest ecosystems. It is estimated that 93% of the original forest cover of these Highlands, above 1500m altitude has been lost (Harvey, Y. *et al.*, 2004).

The race to save the remaining 7% started as far back as 1987 when BirdLife International launched a project to conserve the biodiversity of the Kilum Mountain Forest, the highest point in the Bamenda Highlands. In 2000 conservation efforts were intensified by the now defunct Bamenda Highlands Forest Project (BHFP), still a BirdLife initiative that aimed at conserving all the important forest patches in the Highlands. Following the demise of the BHFP in 2004, the local NGOs that were collaborating with the project, took over the conservation initiative and shared the responsibility for action in the Region on the basis of divisions, among themselves. ANCO took over the Bui Division which had communal forests at Mbiame, Dom and Mbinon. The Mbiame montane forest conservation project was effectively completed in 2005 when the necessary local structures were put in place, their capacities built and the application file for the acquisition of the community forest status submitted to the Ministry of Forests, Wildlife and Fisheries. All over the Bamenda Highlands, the bottom line conservation problem is the excessive pressure exerted on the ecosystems by growing human and domestic animal populations, manifested in various forms of threats, the leading ones being:

- Encroachment for various purposes including crop farming, cattle rearing, planting of exotic species which often become invasive, with the purpose of converting portions of the forest from public utility to private (use of public land for private purposes, over an extended period of years could result in the land becoming private).

- Frequent bush fires, lit by humans either to hunt, clear farm land, or hasten the growth of fresh grass for cattle.

- Unsustainable management of the forest ecosystems for the harvesting of fuelwood, building poles, bush meat, medicinal plants in particular, in the case of the last-mentioned use, *Prunus africana* (Western Highlands Nature Conservation Network: Report on the Illegal Harvesting of *Prunus africana* in the Kilum Mountain Forest in Oku, 2005).

- The intensive pressure on the ecosystems is caused principally by poverty. It is stated that in poverty ranking, people living in the Bamenda Highlands Region, commonly known as the North West Region, come second after those in the Eastern Region, with 52.5% of its population living below the poverty line (Ingram, V. *et al*, (2007). Networks and Networking in the Cameroon (Western) Highlands). If people in the North West are poor, those who live in enclaved rural areas like Dom, are the poorest of the poor.

THE VILLAGE OF DOM

Among the many important forest patches still surviving in the Bamenda Highlands, there is one which is now known to be impressively rich in flora and fauna near the sleepy village of Dom that nestles on the western slopes of Kejodzam range, the northern extension of the Kilum Mountain in Oku. To conserve as well as expand this small forest remnant (452.7Ha) which has been declared to contain "numerous Red Data species that are new to science and restricted to the Cameroon Highlands" (Cheek, M. *et al.,* 2000) ANCO, the Apiculture and Nature Conservation Organisation, and her NGO partners launched early 2008, a replica of the Mbiame Project entitled, *"Integrating biodiversity conservation with income generation for improved livelihood in the Bamenda Highlands of Cameroon".* The objective of the project, like that of Mbiame village before it, is two fold: to reduce the pressure on the fragile submontane ecosystem of Dom in order to ensure its continued existence and fight rural poverty through improving the skills of the Dom people in sustainable farming systems and income generating activities, the so called survival skills (ANCO (2007). Report on Baseline Study of Dom Village).

THE PROJECT

The Dom Project like its counterpart executed in Mbiame between 2000 and 2005, has four components:

- biodiversity conservation, which is the focus of all activities being undertaken

- income generation for livelihood, which targets the building of capacities in activities that generate income and are either supportive of conservation or which are not in direct contradiction to it

- implementation of sustainable farming systems that are either complementary to conservation or that induce permanent or longer cultivation of the same piece of land, thus reducing demand for more farmland

- development of an environmental studies programme for schools in and around Dom village.

BIODIVERSITY CONSERVATION COMPONENT

The activities being executed under this component are the provision of fire breaks, the eradication of the invasive species and the reforestation of the degraded forest areas. Fire breaks along areas from which fire can reach the forest are cleared in November by the community. In the Mbiame case, the project reinforced the efficiency of cleared trails by planting sisal hemp *(Agave sisalana)* along them but in Dom, bananas were preferred because while they are as effective as sisal hemp in stopping fires, they also provide food for some wildlife.

The exotic invasive species stealthily planted in the communal forest of Dom in order to convert portions of it to private ownership is a species of the *Eucalyptus* known as *Eucalyptus kamerunica* [*Eucalyptus* sp. of Dom, in the main plant list]. These will be destroyed by cutting the trunks very close to the ground followed by working mounts over the short stumps to suffocate and destroy them. Vigilance will have to be exercised to ensure that germinated seeds are destroyed as soon as possible. The destruction of the *Eucalyptus* is subsequently followed by planting native tree species in the empty spaces.

Some areas in the former forest are abandoned farm land. Such degraded spaces are being planted with native tree species. The seedlings for planting in the degraded areas are either wildings collected from the forest or seedlings raised by villagers from locally collected seeds and sold to the project. In the latter case, the supplier is paid for both the seedlings and the labour to transplant them.

INCOME GENERATING COMPONENT

This component targets capacity building of Dom community members in three specific areas: apiculture, cane rat rearing and market gardening. In the apicultural sector which ANCO prefers to call bee-farming in order to communicate well with our target group – the rural poor. The organisation carries out capacity building in bee-farming for anybody who so wishes. From among the very many so trained a few serious ones are selected and retrained as trainers for further work back in their villages. ANCO also produces and distributes apicultural materials such as bee-suits, smokers, bee-hives and honey harvesting containers. Assistance with such materials is limited to women and youths who for generations untold, have been kept out of apiculture by local taboos. It was not until 1992 when ANCO, then called the North West Bee-farmers' Association (NOWEBA), introduced training in bee-farming for everyone that women and youths in the Region, after a long period of sensitisation, began timidly to participate in such activities. Now, women and youth participation has increased to the extent that of the over 6 000 persons ANCO has trained since 1994, over 2000 are women and around 1200 are youths. The organisation has also produced and distributed 3575 bee-hives, 2500 smokers, 2680 bee-suits and 4500 honey harvesting and storage 20-litre plastic containers. Thanks to these material inputs, honey production has jumped from a few kilogrammes in 1992 to over 200 tons in 2008, and by the close of the 1990s, had raised household income of those concerned by an average of 11.3% (Anye, M. *et al.* (1997). Sustainable Bee-farming Project, Expost Evaluation).

In order to dissuade village hunters from hunting in the village community forest, they are provided an alternative in training to rear and sell cane rats. During training, the trainees participate in constructing the structure to house the cane rats. In Dom, ANCO began with 10 females and 2 males in the supply centre and are subsequently supplying the trainees with 3 cane rats each (1 male, 2 females). It is hoped that this action will eventually produce the impact ANCO anticipates – to relieve pressure on the now scanty fauna of the community forest.

The market gardening sector targets mostly women and youths who often in the past encroached into the community forests to have access to the fertile forest land for the cultivation of vegetables for sale. Those who are invited for training are those who show proof of having suitable land for cultivating the vegetables. After the training, only those women who have prepared their gardens are assisted with seeds of their choice, for example huckleberry, cabbage, tomato, carrot, spinach, onions, garlic, *etc.* A pool of garden tools has been supplied to assist those who wish to operate their gardens all the year around. A first follow up review showed that three out of the first lot of 21 trainees were successfully harvesting vegetables and spices from their gardens and selling them in the Lasin market, the biggest sub-divisional market near the village of Dom. The others did not completely fail, 12 successfully produced their vegetables and species but sold or ate everything without preserving seeds. Now, they are stuck. They will be assisted if they will provide at least a quarter of the cost for the seeds.

SUSTAINABLE FARMING SYSTEMS COMPONENT

Among the many forms of sustainable farming systems, ANCO has selected 4 which seem suitable in its area of intervention. These are *night paddock manuring system (NPMS), agroforestry, erosion control cultivation, improved fallow, and improved pasture.* Of these systems, the NPMS attracts a lot of attention from the trainees for it tries to create mutually beneficial cross cultural linkages between two races that have for long stayed aloof of each other. Briefly, neighbouring graziers usually from an immigrant race locally called Fulani or Mbororos and the native men or women crop farmers are selected for training which includes an exchange visit to places already practicing NPMS type of farming. Part of the training involves the construction of a fence to enclose a selected farm plot and a *gainako* or cowboy's hut near by. This will be used as a demonstration plot. The next stage is to lead the trainees to negotiate conditions for hiring cattle to stay on the enclosed farm for an agreed duration so as to fertilise it with urine and dung. In some cases, the two parties agree to share the crops to be harvested in a given proportion or may settle for payment in cash. The next element to be agreed on is how the *gainako* would be fed. Usually, it is the farmer who takes responsibility for the feeding. When the negotiation is successful, and both parties fulfill their own parts of the bargain, the system does not only lead to economic benefits to both parties but also to a social fallout which in the Highlands context of sometimes bloody rivalry between the two races, seems much more beneficial than the obvious economic one.

The feeling that trees and crops are "strange bed fellows" is deep rooted in most rural communities of the North West Region. Although ANCO trained 30 farmers and set up a model alley cropping plot and assisted the trained farmers with suitable seedlings in order to encourage them practice the system, follow up studies indicates silent unwillingness to practice the system. It is hoped that when the demonstration plot will eventually start to show the benefits that some of the trainees will be convinced to start to practise the methods learned.

For erosion control, cultivation farmers are trained to use the A-frame to plot out contours and work their ridges across the slopes. They also learn to reinforce these ridges or terraces where the slope is steep through contour bunding, and using the bunds to grow fruits. For the improved fallow system, *Sesbania* or *Tephrosia vogelii* are used on fallow farms for two consecutive years before the farm is worked again for crops. For improved pasture, graziers are trained to plant and tend on their individual grazing land *Tripsacum fasciculatum* (Guatemala grass) and *Brachiaria* species, which will provide cattle feed during the season of fodder scarcity.

ENVIRONMENTAL STUDIES PROGRAMME FOR SCHOOLS

There are two secondary schools in the vicinity of Dom village. These are the Government High School and the Government Technical School both at Nkor, the capital of the Noni Subdivision, Dom is hardly 5km away from Nkor. The environmental studies (ES) programme contains 5 broad types of activities that were planned to be carried out in 24 months as follows:

- Workshops with volunteer staff of both schools to build up an environmental studies programme that fits into the overall programme of the schools. During the workshops ANCO and the teachers select topics and programme them as an environmental study programme for both schools.

- Formulation of a plan to create clubs in both schools. The plan assigns a project technician to a club to work hand in hand with the volunteer staff of the school.

- Incorporation of periodic environmental study activities in clubs; consisting of:

  o Six study visits to the community forest per year

  o School yard decoration by planting ornamental trees and shrubs, wind breaks and shade trees

  o Guest speakers on environmental issues

  o Observation of International Environment Day

IMPLEMENTATION STRATEGY

The first activity undertaken after receiving a request for help, whether written or verbal, is to visit the community and try to understand properly what it is they want. Since most of ANCO's projects deal with community forest conservation or watershed protection, the organisation also visits the site of the intended project. The next step is to carry out a brief survey of the community during which 10 to 15 persons are selected to work with ANCO to draw up a plan for sensitisation.

Sensitisation and raising of awareness (education, information provision) is usually spread over 3 months of periodic visits to the community and entails detailed description of the project activities and the part to be played by the members of the community. Sensitisation should start at least, at the level of the neighbourhood. In the case of Dom, a comparatively small village (2500 people) we visited each household, discussed the project with the family and invited them to the neigbourhood meeting at which further sensitisation was provided. Neighbourhood meetings select at least 10 persons that will represent them at the village assembly. If the project is subsequently funded, this body will meet to elect the Forest Management Institution (FMI) the body that will work with the technicians to implement the project. Also, at the end of each year, the village assembly meets to debate the report presented to it by the FMI. At the end of all this, it could be that the project is not funded. Since this activity costs between $1250 and $2500 US to realise, it is necessary to be cautious in venturing into the process. If seen as a good project, selected ANCO technicians work with the sensitisation team to draft the outlines of the project proposal. In the proposal, the cost of the sensitisation process is budgeted as ANCO's and the community's contribution to the realisation of the project.

The Dom Community Forest Project has made provisions for 12 workshops spread over 24 months. Each of the workshops ends with the construction of a demonstration plot, the *rationale* being that what technical trainers cannot accomplish practically, the workshop organisers have no moral right to expect the local people to succeed in it.

Also, seven of the demonstration plots out of the total that have been established so far are spread all over the village so as to entrench the feeling of ownership of the project which is essential for total commitment. That feeling is further strengthened during the two monthly review and planning meetings which hold under the chair of the President of the FMI. Technicians with their FMI partners present their reports. The subsequent debate may recommend corrections or complete repeat of an activity or a vote of thanks to the executing team. Then follows the planning for the next 2 months when priority is given to activities carried over from the preceding two months. It has already been mentioned that this two monthly meetings are rotative, from one neighbourhood to another.

CONCLUSION AND RECOMMENDATIONS

From her experience in carrying out several community based conservation projects, ANCO finds that such projects rest on five principal pillars: significant community involvement, good governance, income generation, community based leadership in decision making and project execution, and equitable gender representation.

1. Significant community involvement is the first among equals and is the product of intense sensitisation. Such involvement entails the participation of the community at large, the local institutions whether administrative or traditional and any other stakeholder whether supporting or opposing. Failure of a good community based project is an indicator that the sensitisation aspect was either absent or poorly done.

2. Adherence to the principle of good governance, namely: democracy, transparency accountability and equity is the second pillar. Structures set up such as the quarter meeting, village assembly and the FMI should be organised in a democratic manner, preferably through moderated elections to ensure that the resultant structures are all inclusive so that marginalised groups in the community are equitably represented. Democracy here also means bottom-up decision taking and shared responsibilities. To be fully transparent, the owners of the project should actively participate in the proposal preparation including budgeting. Being accountable in this context implies periodic reporting by the FMI to the village assembly at least once a year and thus providing the community the opportunity for full participation and inputs into the project execution process. During such participation, villagers often amply demonstrate that poverty does not deprive them of the ability to make useful contributions to issues they fully understand. Finally, equity is imperative in the distribution of the community forest benefits to community members and in representation at various structures. Good governance is essential for proactive engagement by the whole community.

3. The third pillar is income generation made possible through capacity building. This will ensure that the livelihood needs of the community improve, and will act as an incentive for local people to contribute to the success of the project and ensure its sustainability. As indicated earlier, ANCO does not stop at workshops in the capacity building activity. The new knowledge, concepts and techniques acquired during the workshop must be put into practice to generate a product of high quality.

4. The fourth pillar is community based leadership in decision making and project execution. It is acknowledged that at all stages of the project, the community should be seen to be on the driver's seat. At neighbourhood

meetings, village assemblies, FMI meetings, planning and review meetings a community leader should be in the chair. At the start, the elected leader may be shy and reluctant given that in most villages, an elected leader is never the type of person to preside over meetings. However, he/she should be persuaded to chair all the meetings of the project. This will help to emphasise the village's ownership of the project, for when ownership is blurred or seen to be external, support for the project suffers.

5. The fifth and last pillar is gender equity. To ensure such equity, the project may advise the village assembly to conduct elections at three levels: for men, women and the youths after agreeing on representation quota. This of course should be done when direct voting cannot produce the required results.

6. It is obvious that the weakness of one pillar, and worst of all absence of related considerations, will render the project unstable or unsustainable, and could easily result in its failure. Projects administered by the Apiculture and Nature Conservation Organisation (ANCO), have succeeded in several places because it has stringently adhered to these, above-mentioned five pillars. In the course of this process, ANCO has improved its skills in designing and implementation of projects in community natural resources management.

ACKNOWLEDGEMENTS

ANCO and partners are profoundly obliged to the IUCN National Committee of The Netherlands (IUCN-NL) for the financial and technical support which enabled them carry out integrated conservation projects first in the villages of Mbiame and currently in that of Dom. For further information, please visit ANCO's website at: www.anco-cameroon.org.

Published in *The Nigerian Field* 72: 101-107 (2007) among papers presented at the NFS [Nigerian Field Society] symposium on 'Conservation in South Eastern Nigeria and Cameroon' held at the Royal Botanic Gardens, Kew 2007.

TOWARDS A CHECKLIST OF THE THREATENED FOREST REMANTS OF DOM, BAMENDA HIGHLANDS, CAMEROON

Yvette Harvey

Herbarium, Royal Botanic Gardens, Kew, Richmond, Surrey, TW9 3AE, UK

Summary. This presentation seeks to present work towards a conservation checklist of the Community Forest of Dom, Bamenda Highlands, North West Province, Cameroon. The forests are classed as submontane and include a number of threatened plants, including the critically endangered *Newtonia camerunensis*.

The "Wet Tropics Africa Team" of the Royal Botanic Gardens, Kew has been working in collaboration with the National Herbarium of Cameroon for over ten years to produce Conservation Checklists for botanically important or threatened areas of Cameroon. The Mount Cameroon conservation checklist (1998) has been our model. Mount Cameroon is one of the few places where natural vegetation exists from the coast to sub-alpine elevations and includes one of only three locations where annual rainfall exceeds 10 m per annum. So far five checklists have been produced and the Darwin Initiative has provided funding for a further three, one of which is to be a Checklist of the Community Forest of Dom. We are looking to do similar work in Guinea and Gabon.

Dom is located in the Bamenda Highlands of North West Province, Cameroon, which themselves form part of the Cameroon Highlands. The Cameroon Highlands run from Bioko and Mount Cameroon in the south, northwards to Tchabal Mbabo and then continue eastwards as the Adamoua area, following a geological fault line. It has been estimated that 96.5% of the original frest cover of the Bamenda Highlands above 1500m altitude has been lost (Harvey et al. 2004).

A Bamenda-based NGO called ANCO (Apiculture and Nature Conservation Organisation) requested our help with the forest of DOM in 2004. ANCO is a member of WHINCONET (Western Highlands Nature Conservation Network), a consortium of NGOs. WHINCONET was founded in 2002 and is an organisation recognised by the Cameroon government. Its objectives are to enhance biodiversity conservation and poverty alleviation by sustainable management of natural resources through networking. Kew has been asked, by NGOs within WHINCONET, to survey many of the remaining forest fragments within both North West and South West Provinces. With the guidance of the surveys, WHINCONET will then choose the most interesting areas, in terms of diversity, endemism and numbers of threatened taxa present, to conserve. Owing to insufficient resources, it is not possible to save all areas. The consortium is currently headed by Mr Paul Mzeka who runs ANCO.

ANCO seeks to support local communities protecting their forests. One of their projects is a nursery, of approximately 400m² and during early 2005 this contained an estimated 30,000 plants, including *Prunus africana* and the threatened taxa, *Newtonia camerunensis, Dovyalis cameroonensis* and *Eugenia gilgii* with funds from Kew. Seedlings are transplanted to reforest areas. In fact the Dom community were the recipients of *Prunus* seedlings in June 2005. Seen in Plate 2F is George Kangong, chief nurseryman, in discussion with Marcella Corcoran, one of Kew's specialist horticulturalists. Marcella, in the Spring of 2005, spent time training ANCO staff in tropical horticultural techniques.

Dom is in Noni subdivision (under Bui division) near its capital Nkor. Its forest came to wider attention when the people of Dom appealed to the Bamenda Highlands forest project (since closed) for help in managing the forest, and achieving community status, some years ago. ANCO began taking up this challenge early in 2005 with NC-IUCN assistance. ANCO suggested to Kew that it would be useful to survey the forest, and so three botanical survey visits were made in April/May 2005, February/March 2006 and September 2006, by combined RBG Kew/YA/ANCO teams.

In a community forest, forest management includes both forest conservation activities and sustainable use of natural resources. It is owned and managed by the local population that allow both conservation and sustainable use of the natural resources in the forest (i.e. harvesting of medicinal plants, hunting and bee-keeping). In Cameroon, the law on community forestry dates from 1994 (No. 94/01, 20th January 1994). Management includes keeping an eye on various designated taxa. For plants in submontane areas this will include: *Gnidia glauca* (Plate 7D), this tree is usually seen in regenerating forest and is an indicator of positive changes in the forest; on the other hand, *Pteridium aquilinum* (bracken) (Plate 8G) is seen as an indicator of negative change since it grows in forest open spaces, many of which are created by man-made fire.

Discussions with the Divisional officer in Nkor appear to show a direct link between deforestation (Plates 2A-C) and an erosion of water supplies. Reduction in the dry-season flow of water, which he attributes to forest clearance in Noni, had resulted in 300 women appealing to him for help. It appears that Noni "landgivers" who were traditionally accepted to have the right, had given out tracts of forest to Nso farmers in the North, in return for a tithe on the resultant crops. The Divisional Officer expressed the high value that he placed on forest conservation, not just for watershed protection, but also to conserve rare species that might be of use for future generations. He has been very active in visiting the forest at Dom and in chairing meetings of the forest management committee.

Our collecting teams typically comprise of personnel from the NGO that we work with, in the case of Dom this is Kenneth Tah, staff from the National Herbarium of Cameroon, and guides from the local village. The forest fragments are reached on foot via a network of paths. Forests are generally clinging to very steep, mountainous slopes. Specimens are gathered in polythene bags prior to being placed in a plant press during frequent stops. With the permission of the Fon, a significant number of Noni plant names were gathered, including local uses, it having been agreed with him that these could be incorporated in our proposed checklist. The Fon was particularly keen that this aspect of the traditions of the Noni people would not be lost.

The submontane forest at Dom has the world's largest known population of the critically endangered *Newtonia camerunensis* (Plate 5C). This is a submontane and montane forest tree, known only from Cameroon's Bamenda Highlands and Bamboutos mountains. *Newtonia camerunensis* is threatened by forest clearance for timber, firewood and in the case of Dom, small-scale agriculture. Other critically endangered plants seen were *Eugenia gilgii, Chassalia laikomensis, Psychotria moseskemei, Oxyanthus okuensis* (Plates 6C-D)and *Dombeya ledermannii* (Plate 7C). The endangered *Allophylus conraui* [*Allophylus ujori*] was also encountered.

The edges of the forests were marked by populations of the 2m+ high, white flowered *Lobelia columnaris* (Plate 4A); *Clematis villosa* subsp. *oliveri* (Plate 5F), an erect perennial herb to about 60cm, with solitary white flowers to 5cm diam.; the large leaved *Solanecio mannii* (Plate 4E) , a shrub reaching 7m, that, although widespread in Africa, is quite distinct as there are very few large-leaved Asteraceae [Compositae] occuring in the Cameroon Highlands; the 2m tall bulb, *Drimia altissima* (Plate 8A); another showy bulb, *Scadoxus multiflorus* with a red star-burst style inforescence; and *Gnidia glauca* (see Plate 7D), the 15m high tree with 5cm diam. flower heads.

Within the forests, along with the endangered taxa listed above, were many trees and shrubs, including the showy red-flowered *Kigelia africana* (Plate 3G) whose fruits are shaped like a salami; the 2m shrub, *Psychotria psychotrioides*, (Plate 6E), found in forests throughout much of tropical Africa; also in the Rubiaceae, *Cuviera longiflora* (Plate 6B), a tall 8m shrub that was swarming in ants; the aroid, *Anchomanes difformis* (Plate 7F), the rarer of the two taxa known to occur in Cameroon; and *Tabernaemontana* sp. of Bali Ngemba (Plate 3F), a Cameroon endemic also found within Bali Ngemba, one of our checkist areas, with pleasantly aromatic flowers, many of which carpeted the forest floor.

In the three brief visits made so far, over 350 collections have been made, and numerous sight and photographic records have been taken. It is hoped that the checklist will be published by 2010 as part of a joint Kew and National Herbarium of Cameroon, Darwin Initiative supported project (no. 15034). So far, our expeditions have shown Dom to have the biggest living collection, in terms of numbers present, of the critically endangered *Newtonia camerunensis*, and for this reason alone, it is likely to be one of the forests chosen by WHINCONET for conservation.

As mentioned earlier, the checklist production follows a pattern. There will be several plant collecting trips made to the specific area at different times of the year to capture the flowering and fruiting stages of the plants present. Several duplicates are made of each collection. At the time of collection, all field books are entered into a specimen database and labels are produced. The top-set of specimens will remain at the National Herbarium of Cameroon (this is one of the conditions of our research permit, and agreement between our two institutions). The second set of specimens are sent to Kew where they are identified by family and/or regional specialists. These identifications are subsequently added to the specimen database. Then begins the checklist production phase. The checklists have a number of introductory chapters before the species list. Species accounts are compiled and downloaded from the database (the database has an ever expanding "folder" of species descriptions), and these make the bulk of the checklist. The checklist is compiled in an alphabetical arrangement: species within genera, genera within families and families within the groups Dicotyledonae, Monocotyledonae, Lycopsida (fern allies) and Filicopsida (true ferns). All species listed within this part of the checklist will have an IUCN rating, following IUCN 2001 criteria. An important introductory chapter includes full red-data assessments of the most vulnerable species present in the area, along with management suggestions.

The checklists are distributed within Cameroon to government bodies, local government offices, NGOs, interested parties and local schools. However important the information within, a book is likely to sit on a shelf gathering dust. As a consequence, many people living in the areas may not have access to this printed source. As an effective, yet inexpensive form of publicity, our team prepares posters for plants found to be of high conservation status that occur within the checklist area. They are distributed to government bodies, local government offices, NGOs, and placed in areas where people congregate, such as schools, restaurants, bars and hospitals.

A BRIEF HISTORY OF THE NONI PEOPLE

George Nkwanti
Government High School, Nkor Noni Subdivision, Bamdenda Highlands, Cameroon

The present Noni tribe is an extraction of the Tikar. They are said to have originated from Tikar, an area along the River Mbam in the Central Region of Cameroon. They came alongside the Nso people and made their first settlement with the Nso at Kifum. Following the outbreak of the Jihads or the Fulani Invasion in the 19th century they disintegrated and some of them found refuge at Tabeken in Nkambe while others migrated to Oku. Following internal conflicts two Nso princes broke away and established the chiefdoms of Mbiame and Oku some centuries ago. One faction of the Noni people followed the Oku prince. From Oku, some broke off and moved with princely belongings to found the chiefdoms of Djottin and Din. Another faction of this group moved through Mbim and founded the chiefdom of Dom. These groups of people in Dom today are referred to as the Kih. History reveals that the first settler in Dom was one Nji Kokembang. Even the Kih people who are said to have been the first settlers in Dom met him already there in an area called Nkali where the Dom people first settled.

The Noni group at Tabeken also disintegrated following a succession dispute and moved with royal belongings and founded the present chiefdom of Nkor, Tfuh and Laan. See Plate 3A. Another faction is said to have come through Dumbu to join those from Tabeken in Nkor and Laan. For instance, the Mbin people are said to have moved through Dumbu to Nkor where they founded the present Mbin Fondom following a friendship arrangement with the Fondom of Kochi. The present Kochi people are generally referred to as the Boluh because of their source of origin i.e. people from Luh.

Another group of people later moved from Kikaikelaki and founded the sub chiefdom of Mvunjeiwu which pays allegiance to the Mbin fondom.

These Noni villages are said to have suffered domination by neighbouring tribes. For instance the Tfuh people suffered a lot from Nso domination, Din also suffered Nso domination. The Oku people also tried to dominate Din and Djottin while Nso also dominated Dom and Djottin villages. These Noni villages later fought this domination and had their autonomy.

The Noni people established well organised traditional institutions and culture. At the head of the traditional institutions is the Fon who is closely assisted by a regulatory society called in Noni language gwifon (the Ngumba). It has the power to exile recalcitrant inhabitants of the tribe. Another serious traditional institution is the war society generally known as the nfuh which announces war and fight for the tribe. There are also sacred shrines where they offer sacrifices to the gods. The sacred shrines are found mostly in the heart of the forest where the Noni people believe the gods inhabit.

Though the Noni tradition and culture is uniform in all the villages, there are some slight differences in practice. The Djottin/Din practice of "kati kati" is not practiced in the villages of Dom, Tfuh, Laan and Nkor. This difference is due to inter-tribal influences. For example Din and Djottin villages have a lot of influence from the Kom and Oku people who practice "kati kati" (see Footnote) especially during funeral ceremonies. The language also suffered some slight differences. The accent of a Din person is different for example from that of somebody from Laan. This is still due to influence of neighbouring tribes. The Noni people after establishing their settlements were very hospitable to the extent that they allowed neighbouring tribes to mix freely with them to the extent that the Nso people names were given to Noni villages. An example is the case of the Tfuh-Nso relationship which led to the creation of another name for the Tfuh people as Mbinon. History also reveals that the people of Dom had the name Bvugoy but the Nso people changed it to Dom. Also, Laan became Lasin.

At the level of religion, the Noni man practiced first the traditional religion and later embraced Christianity like the other Tikar tribes.

Footnote: "kati, kati" is a Kom expression meaning to cut into tiny pieces. Among the Highlanders, it is applied mostly to roasted chicken (Paul Mzeka).

HISTORY OF FOREST CONSERVATION IN DOM

Paul Mzeka

Coordinator ANCO, P.O Box 5150, Bamenda, Cameroon.
Tel: +237 77 71 56 5 / 99145918
Email: mzeka.anco@yahoo.com   Website: www.anco-cameroon.org

INTRODUCTION

Though hardly 20km away from Kilum in Oku, the flagship area for forest conservation in the Bamenda Highlands, Dom is so highly isolated that it is not surprising it was never affected by the environmental awareness raising launched by BirdLife International in 1990. It was not until 2002 that technicians from the now defunct Bamenda Highlands Forest Project (BHFP) visited the subdivision and the Fon (chief) of Dom attended the meeting organised to raise awareness on conservation in the area. The Fon reports that no one from the BHFP visited Dom though he had, during the meeting invited them to do so. Before closing down in 2004, the BHFP handed a list of forests they had surveyed to the five NGO's with whom they had been collaborating. The collaborating NGO's distributed the surveyed forests among themselves and Mbinon, that topped the list for being of high conservation interest, was given to ANCO.

PRELIMINARY INVESTIGATION

Our first visit was to Mbinon in the early 2000s, and after a preliminary investigation about continuing with the work of the BHFP, we realised the problem of ownership of Mbinon forest which had come up during the BHFP survey, was intractable. Our investigation also revealed that Mbinon forest was part of a submontane forest that extended to Dom. We therefore decided to visit Dom and were eagerly received by the Dom traditional leaders who expressed their determination to conserve their own portion of the forest which we later found to contain 3 blocks totaling 452.7ha. We took time to inform them that their forest could either continue to be conserved as a communal forest, or the community could request the Senior Divisional Officer to issue a prefectorial order declaring the forest as a conserved area or they could work to request the Minister of Forestry, Wildlife and Fisheries to declare it a Community Forest. We took time to explain the advantages and the disadvantages of each category. Finally, they were asked to choose the category they wanted. After they had chosen to follow the process to obtain the status of a Community Forest for their forest reserve, (Dom CF), we then agreed on the plan to sensitise the whole community during which, the village leaders were to play a leading role. It was also agreed that the awareness raising was to begin from households and to take a total of 8 days, two days per week. After this, household delegates would meet at the level of quarters for intense sensitisation. This was going to take 4 days in two weeks. Finally, the elected quarter delegates were to meet twice at the village assembly to select a committee to work with ANCO technicians to write up the project proposal. The village assembly was also to choose the broad activities that the project should target. After the discussion and agreement, the time to start the awareness raising was agreed on and we left.

MEETING WITH THE COMMUNITY

The meeting with the community was, as has already been hinted, a process that began with contacting households and explaining the implications of a community forest and the role of the community members in its realisation. Each household chose 2 or 3 of its members to the quarter meeting. ANCO took the opportunity to concurrently carry out a baseline survey of the community. At the quarter meeting, members were assisted with prepared questions to discuss the main issues of Dom CF. Each quarter had two meetings chaired by an elected chairman. During the second meeting, quarters elected their delegates to the village assembly according to their population size.

The Dom village assembly was given three main duties:

- to elect a committee to work with ANCO technicians to draft the project proposal whose major activities were proposed by the assembly. The proposal when ready was presented to the assembly.

- to elect the Forest Management Institution (FMI) to work with the technicians in the execution of the project.

- to meet once each year to debate the report of the project execution presented to it by the FMI President. The project approved by the village assembly is then fine tuned by ANCO technicians and copies sent to a number of funding bodies. The expenditure incurred in carrying out the awareness raising activity is budgeted in the project proposal as ANCO's and the Community's contribution to the realisation of the project.

METHODOLOGY OF PROJECT EXECUTION IN DOM

The first activity we undertook after receiving a request for help, from Dom was to visit the community and try to understand properly what it was they wanted. We also on this occasion, visited the site of the intended project. The next step was to carry out a brief survey of the community during which 10 to 15 persons were selected to work with us to draw up a plan for awareness raising.

Our awareness raising was spread over 3 months of periodic visits to the community and entailed detailed explanation of the project activities and the part to be played by the members of the community. Dom is a comparatively small village (2500 people approx.) so we visited each household, discussed the project with the family and invited them to the quarter meeting at which further awareness raising was carried out. Quarter meetings selected at least 10 persons that represent them at the village assembly. The assembly elected the Forest Management Institution (FMI). Since the awareness raising costs between $1250 and $2500 US to realise, one has to be cautious in venturing into the process. In the proposal, the cost of the awareness raising process was budgeted as ANCO's and the community's contribution to the realisation of the project.

The Dom CF project contains 12 workshops spread over 24 months. Each of our workshops ends with the construction of a demonstration plot for we reason that what we cannot as technicians, successfully accomplish practically, we have no moral right to expect the trainees to do so.

Also, seven of our demonstration plots out of the total were planned to be spread all over the village in the quarters so as to entrench the feeling of ownership of the project which is essential for total commitment. However, in the process, the idea was abandoned due to high cost. Two monthly review and planning meetings are held under the chair of the President of the FMI. Technicians with their FMI partners present their reports. The subsequent debate may recommend corrections or complete repeat of an activity or a vote of thanks to the executing team. Then follows the planning for the next 2 months when priority is given to activities carried over from the preceding two months. We have already mentioned that this two monthly activity is rotative, from quarter to quarter.

Tree species ANCO has often nursed to carry out restoration ecology either in a community forest or watershed consist of any native species that can be regenerated either through nursing their seeds or by cuttings. Also, some exotic species are nursed to meet the demand of agroforestry farmers. Some of the commonest ones nursed and transplanted in both Kitiwum and Dom are listed below.

Genus/ Species	2007		2008	
	Nursed	Transplanted	Nursed	Transplanted
Prunus africana	10710	1255	5245	4667
Ilex mitis	2750	2621	4650	3995
Calliandra sp.	2030	2005	3475	3250
Croton macrostachyus	550	450	6641	5676
Milletia sp.	2800	2667	4660	4120
Albizia gummifera	1240	1175	3255	3007
Jacaranda sp.	2100	2050	6941	6800
Cassia sp.	820	768	5645	5514
Leucaena sp.	3465	3245	2150	2050
Acacia sp.	2100	2001	1140	1010
Pittosporum viridiflorum	540	5251	6750	6525
Oncoba sp. nov. *	757	696	5547	5341
Eugenia gilgii *	600	594	3641	3344
Dovyalis cameroonensis *	325	322	3670	3250
Total	**31,067**	**25,380**	**64865**	**59,954**

Among those on the list, the endangered species marked (∗) have been successfully nursed and transplanted with technical assistance from the Royal Botanic Gardens, Kew and financial assistance from the Bentham-Moxon Trust (managed at Kew). On the whole, our success growth rate has been 70%. This high rate is as a result of nursing the seedlings in polythene pots for almost 6 months before transplanting. To ensure that the seedlings are ready for transplanting before end of August, the nursery work must begin in January so that by August ending, transplanting has been completed.

FOREST RESTORATION IN DOM

Kenneth Tah

ANCO, P.O Box 5150, Bamenda, Cameroon

Over the past few decades the forest in Dom witnessed a drastic reduction in size because of intensive human activities in the area. The quest for fertile farmland, pasture and the positive reaction from traditional authorities facilitated the destruction of thousands of hectares of forest. The traditional setup of the Noni clan and the greediness and shortsightedness of some notables favoured the wanton destruction. Notables who are regarded as the landowners or custodians of the land commonly called "landlords" have over the years leased out huge portions of forest for farming and grazing to mostly non-indigenes of Dom. This irresponsible practice unfortunately gained popularity among the notables in the Noni area and has been the major cause of forest destruction till present.

In the early Nineties some attempts were however made by some of the notables to ensure some reforestation of the areas under cultivation. The tradition and customs of the area do not allow for deliberately uprooting anything planted conscientiously. The notables therefore obliged farmers to plant *Prunus* trees in the portions they were cultivating. Today some stands of *Prunus* are visible in the farms. This approach however, was faced with some shortcomings because no mechanism was put in place to ensure that the farmers actually planted trees and provide adequate care. The deforestation of the forest continued without any resistance from the authorities in place.

In 2005, after the exit of the Bamenda Highlands Forest Project which showed some interest in working with the people of Dom, ANCO received funding from IUCN Netherlands to help the Dom community protect the remaining patches of forest as community forest. Top on the list of activities to be implemented was the conservation of what still persists and the restoration of the degraded areas. There exist three patches of fragmented forest covering an area of 452.7ha. The reforestation had as a main goal the maintenance of the three patches while closing up the corridors that exist between them.

ANCO'S APPROACH

Building from her experiences working with other communities, ANCO decided to actively involve the entire community at all stages of the restoration process. The approach provided short and long-term benefits to the community. Community members were trained in seed and wildings collection, nursery establishment and care. Demonstration plots were created to serve as models for the rest of the community to emulate. Individuals established private nurseries with seeds and wildings collected from the forest. ANCO provided pots and additional seeds of threatened native species for introduction into the nurseries. During the planting season, the seedlings are bought by ANCO and transported to the forest where each individual is responsible for the planting and care of the seedlings that they sold. It should be noted that these activities are regarded as family focused and all members of the family are encouraged to take part in the process. The immediate benefit of cash payment encourages active participation.

Different seed storage and treatment techniques are used incorporating indigenous knowledge to ensure viability and high performance of seedlings. Common seed treatment techniques based on local experience are used for the seeds as well as trial and error especially for those species with no known standard seed treatment techniques. The common techniques include; heat treatment, exposure to sunlight and water treatment.

The planting distances vary with the different species but generally a planting distance of 2-3m is respected. The survival rate in the field has been very varied depending on the source of germplasm. Seedlings planted from wildings have a relatively poor success rate than those from seeds nurtured in the nurseries. On the other hand it is easier growing a tree like *Newtonia camerunensis* from the wildings than from the seeds because of the very low germination rate of the seeds. Over 10,000 rare plants have been planted so far in Dom with the assistance of the Royal Botanic Gardens, Kew. The current phase of the project which started in 2008 will see the planting of 30,000 more seedlings over an area of 60ha of degraded land. Enrichment planting will also continue with the addition of more rare species.

CHALLENGES

There have been some challenges in the restoration of the Dom forest. Bushfire and grazing have led to the destruction of some trees planted in 2005. Storage of seeds has been a major setback as some of the seeds are eaten by insects and small rodents during storage. For example most of the fruits of *Dovyalis* spp. were eaten up by rats a few days after collection. More trials and propagation techniques needs to be done on the rare plants to know the most appropriate techniques to treat and nurse their seeds. For more information on ANCO, visit our website at www.anco-cameroon.org

INTRODUCING ANCO

Paul N. Mzeka
Coordinator ANCO, P.O Box 5150, Bamenda, Cameroon.
Tel: +237 77 71 56 5 / 99145918
Email: mzeka.anco@yahoo.com   Website: www.anco-cameroon.org

ANCO (Apiculture and Nature Conservation) was started in 1992 and legalised the following year as the North West Bee-farmers Association (NOWEBA) to train rural communities living in and around conserved areas (community forests, forest reserves, watersheds etc) to sustainably exploit them for livelihood through beefarming. The central purpose was to fight rural poverty while at the same time trying to convert such rural communities to conservation. In pursuing the latter, beefarmers were trained and assisted to plant melliferous tree species. By 2000, NOWEBA had trained over five thousand beefarmers and assisted them with various quantities of beefarming tools and material. They had together planted a total of 60,195 melliferous trees.

The quantity of honey and beeswax being produced (50 tons and 1 ton per annum respectively) needed a marketing structure and facilities that ANCO was unable to provide. Thus in 2000, the organisation was restructured to create a component for marketing (Honey Cooperative) and the other, as a specialised organ for capacity building in apiculture and nature conservation (ANCO). Since 2000, ANCO has actively participated in the protection and reforestation of 3 community forests, 11 watersheds and one bat sanctuary. It has also trained over 2600 farmers in Sustainable Land Management (SLM).

ANCO integrates conservation with poverty alleviation. In conservation, the main focus, she carries out restoration ecology, bush fire management, forest inventory and sustainable exploitation of non-timber forest products (NTFPs). For poverty alleviation, communities are trained in apiculture, market gardening and cane rat rearing and in SLM: agroforestry, night paddock manuring system, improved fallow, improved pasture and erosion controlled cultivation.

ANCO has received financial and technical support both nationally and internationally. Nationally from the British High Commission in Cameroon, the Swiss Development Organisation (HELVETAS) Cameroon, the Dutch and German Embassies, the German Development Organisation (DED), the Canadian High Commission in Cameroon and the Government of Cameroon (Ministry of Forest, Wildlife and Fisheries). Internationally: the IUCN National Committee of The Netherlands (IUCN-NL), Village Assistance in Development UK, the Royal Botanic Gardens, Kew, International Tree Foundation UK and the Virginia Gildersleeve International Fund USA.

NGOs are reducing their active participation in the conservation of forest patches as communities develop skills in managing them. ANCO plans to turn more of her conservation attention to restoration ecology. For more information, visit our website www.anco-cameroon.org.

REFORESTATION WORK AT ANCO

Paul Mzeka
Coordinator ANCO, P.O Box 5150, Bamenda, Cameroon.
Tel: +237 77 71 56 5 / 99145918
Email: mzeka.anco@yahoo.com   Website: www.anco-cameroon.org

INTRODUCTION

Reforestation at ANCO is carried out as an integral part of conservation which focuses on forest patches, remnants of the once extensive submontane forest that not long ago, covered more than 60% of the Bamenda Highlands. Most of the patches are being conserved either as community forests or to protect watersheds. Conservation in either case involves the protection of what is and its enhancement through the planting of indigenous species. The protection of what still survives calls for the prevention of bush fires and the stoppage of encroachment either for farming, poaching or cattle rearing. It also includes, the elimination of invasive species mostly the *Eucalyptus*.

ENHANCEMENT THROUGH TREE PLANTING

In general, most of the community forest patches and watersheds in the Highlands have been deforested, leading in the case of watersheds, to pipe silting and reduction in water volume. Deforested areas are restored through protective measures, enriched by tree planting. Areas from which invasive species have been eliminated are also restored through tree planting

To produce seedlings for eco-restoration, ANCO works with the beneficiary communities to set up what has been called central nurseries. These are huge nurseries established in one spot and containing about 50,000 seedlings or more. Two or three village youths are engaged to care for them (weed, water and manage pests and diseases) until the seeds are transplanted. Using this approach and with technical and financial help from IUCN Netherlands (IUCN-NL) we succeeded to nurse and transplant over 75,000 trees in the Mbiame Community Forest between 2000 and 2004. Between 2004 and 2007, with help from the International Tree Foundation UK and the Royal Botanic Gardens, Kew, we established a vast nursery in the village of Kitiwum to reforest their barren watershed (see Plates 2E-F). Assistance from Kew was specifically used to nurse 4 threatened species: *Newtonia camerunensis, Oncoba* sp. nov., *Eugenia gilgii* and *Dovyalis cameroonensis*.

For our on-going project in the village of Dom we have abandoned the unique nursery approach in favour of the decentralised approach. We began by asking for volunteers to participate in a 3 day workshop to develop their skills in nursery. We had 27 persons of both sexes. After the training, the participants were assisted with seeds mostly of threatened species. They were told that all the seedlings they successfully grow will be bought by the project. The price per seedling was negotiated and we later discovered that this is by far, a better approach to seedling production than the first one for the following reasons:

- The target quantity is easily realised.
- Community members happily participate because they see their personal benefit involved.

- Nursery skills are widely spread in the community.
- Because of their personal gains, participation in transplanting is high. We are still to find out whether care for the planted trees will also improve.

As concerns transplanting of the seedlings, we do not support the approach of mobilising the village to plant the trees for free unless it is their expressed wish to contribute in that manner. Our view is that if the community has sacrificed their land for the forest, they should not be expected to sacrifice more. Forest conservation should not become a drain on the community's scarce resources of time and money but should be seen as contributing to the reduction of poverty in the concerned community.

CONCLUSION

It has been estimated that over 93% of the original forest cover in the highlands above 1500m altitude had been destroyed (Harvey *et al.* 2004). There is of course urgent need to conserve the tiny patches that still remain, however there is greater need to carry out restoration ecology not only to contribute to climate change but also to reduce the deterioration of the highlands environment.

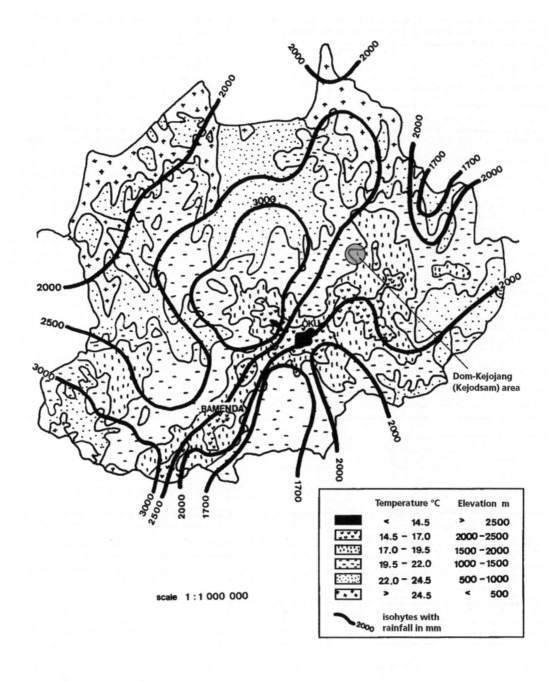

Fig. 5. Rainfall at Dom-Kejojang. Reproduced with the permission of BirdLife International from McLeod (1987).

GEOLOGY, GEOMORPHOLOGY, SOILS & CLIMATE

Martin Cheek

Herbarium, Royal Botanic Gardens, Kew, Richmond, Surrey, TW9 3AB, UK

Dom and the Kejojang forests are situated on a west-facing protrusion of the High Lava Plateau. Below them is the steep scarp slope that connects the High Lava Plateau (1500–2100m alt.) of the Bamenda Highlands with the Low Lava surface (900–1200m alt.) on which the former appears to rest; see Fig. 1 (Hawkins & Brunt 1965). The basalts and trachytes that form the bulk of the Bamenda Highlands are of ancient volcanic origin (Courade 1974). The Bamenda Highlands themselves are the most extensive of the mountainous areas, usually known as the Cameroon Highlands, that begin with the Atlantic Ocean islands of São Tomé, Príncipe, Annobon and Bioko and proceed NE, in a band 50–100km wide. These mountains have their origin in a geological fault and are formed largely of igneous material. Three main periods of volcanic activity and one of plutonic uplift have been reviewed by Courade (1974). The Bamenda Highlands were largely formed in the Tertiary, in the second of the three main periods of volcanic activity 'the middle white series'. However Courade classifies the geology of the Dom area as Pre-Cambrian basement complex. If this is not a mapping error, then Dom and its surroundings have been uplifted by tectonic activity.

Soils are classified as humid volcanic soils of the High Lava Plateau by Hawkins & Brunt (1965), and a volcanic crater is indicated in the neighbourhood of Dom, although this has not been located by us on the IGN 1:200,000 Nkambe map. Courade classifies the soils of the Dom area as being medium fertility, third class (out of four classes) in terms of fertility. Tye (1986) refers to soils in the Bamenda Highlands area as being 'clayey', resulting in more permanent streams than some of the mountain areas of more recent origin in the Cameroon Highlands, such as Mt Kupe and Mwanenguba.

Courade's rainfall map shows the area in which Dom and Kejojang falls being somewhere between the 3m and 4m isohyets as is the Mt Oku (Kilum-Ijim) area. However, this may be a mapping error since other sources (see below) suggest a lower figure. No meteorological station has been located in the area around Dom insofar as we have been able to find data, although one might be expected at Nkambe.

The general climate of North West Province is broadly the same as that of the rest of West Africa (Hawkins & Brunt 1965) with a rainy season between May and September and a dry season between October and April.

The mountainous nature of the Bamenda Highlands alters this general pattern somewhat. Main rain-bearing winds coming from the south-west are interrupted by high land, causing very wet south-west facing slopes on the mountains such as in the Dom-Kejojang area and resulting in rain shadows on the north-east sides. Temperature inversions occur in valleys and depressions. The drainage of cold night air into lower areas results in lower than expected temperatures and thick mists in the foothills. During the day, warm air ascends and leads to cloud formation and misty conditions along escarpments. The intensity of the Harmattan, a dry north-easterly wind which may be heavily laden with dust, is also influential in determining weather conditions. Having a westerly aspect, the forests of Dom can be expected to be sheltered from the worst of the Harmattan in the dry season.

The combination of altitude, temperature inversion, slope orientation, harmattan, mist and cloud leads to the development of different local climatic zones (Hawkins & Brunt 1965). Nine climatic zones are defined by Hawkins & Brunt (1965) for the whole of North West Province, that applying to the Dom area (and also the neighbouring Ndu tea estate and Mt Oku) is Zone 7: "Cool and Misty". The characteristics of this zone are as follows (extracted from Hawkins & Brunt, 1965).

Temperature. Mean maximum: 20–22°C; mean minimum: 13–14°C. November has the lowest mean minimum temperature and December the highest mean maximum. Temperature inversions at night in narrow valleys which suffer from bad air drainage leads to some ground frost, mainly in January or February.

Rainfall. Varies from 1780–2290mm per year (average annual rainfall for Jakiri is 2000mm). See Fig. 5. Most rain falls between July and September. The summit of Mt Oku leads to drier conditions to the east of the mountain.

Humidity. Generally January and February have the lowest relative humidity (average 45–52%). The monthly average exceeds 80% in July and August (Bambui Agricultural Station). During the rainy season, mist and low cloud occur frequently.

VEGETATION

Martin Cheek

Herbarium, Royal Botanic Gardens, Kew, Richmond, Surrey, TW9 3AE, UK

In this chapter vegetation encountered thus far in the ara of the Dom forest is discussed by references to that classified elsewhere in the submontane belt of the Bamenda Highlands, that is at Bali Ngemba Forest Reserve (Cheek pp. 13-26 in Harvey *et al.* 2004).

Four of the five vegetation types recognised at Bali Ngemba, and several of their subdivisions were found and studied at Dom, as follows:

1. Submontane forest (1300 to 1900m alt.)
    a) *Lower submontane forest with Pterygota (1300-1500m alt.) was not found at Dom*
    b) Upper submontane forest with *Pterygota,* and forest edge (1500-1700m. alt.)
    c) Extreme upper submontane forest (1700 to 1900-2000m alt.)

2. Montane forest, grassland and scrub-forest edge (1900-2000 to 2200m alt.)
    a) Montane forest. Incompletely documented in Dom
    b) Montane grassland and scrub-forest edge

3. Submontane inselberg grassland and rock faces (1600-1930m alt.)
    a) Rock faces
    b) Inselberg grassland (see 4. Derived Savanna)

4. Derived Savanna (1600m alt.)

Each of these vegetation types is illustrated with case studies from sites at Dom. However, since our exploration of Dom has been so incomplete, these vegetation types have not been mapped. For each vegetation type physiognomy and species composition is examined, sampling levels are gauged, endemic threatened taxa are indicated, threats are documented and phytogeographical links are assessed.

Since our studies at Dom have not been exhaustive it is likely that further studies will find addional species in each vegetation type.

PREVIOUS STUDIES

As was the case at Bali Ngemba (Cheek in Harvey *et al.* 2004), no on-the-ground studies of vegetational composition at Dom appear to have been conducted prior to our own.

One previous provincial (Hawkins & Brunt 1965) and one national (Letouzey 1985) vegetation map of Cameroon show the forest patch corresponding to Dom. The relevant parts of both these maps are reproduced in this book as Fig. 6 and Fig. 7 respectively. Although the compilers of both maps did extensive botanical ground-truthing in the progress of their work, neither actually visited the Dom area, as far as is known, but characterised the area of the forest from aerial photography and its vegetation type by extrapolation from observations made elsewhere in the Bamenda Highlands.

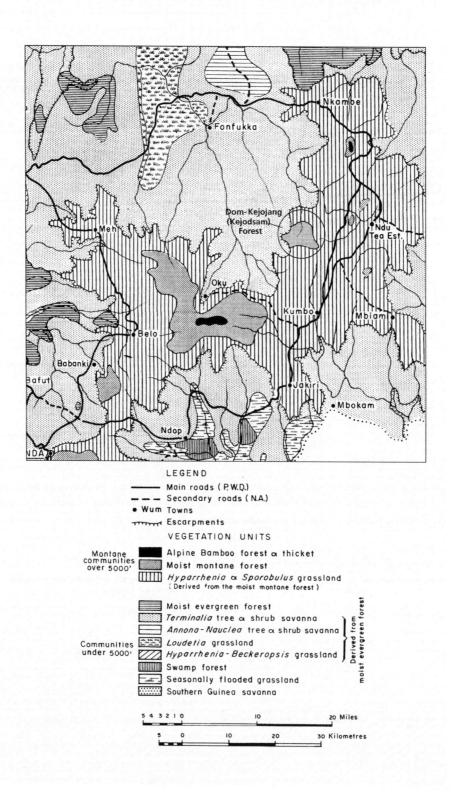

LEGEND

— Main roads (P.W.D.)
- - - Secondary roads (N.A.)
• Wum Towns
↑↑↑↑↑ Escarpments

VEGETATION UNITS

Montane communities over 5000'
- Alpine Bamboo forest α thicket
- Moist montane forest
- Hyparrhenia α Sporobulus grassland (Derived from the moist montane forest)

Communities under 5000'
- Moist evergreen forest
- Terminalia tree α shrub savanna
- Annona - Nauclea tree α shrub savanna
- Loudetia grassland
- Hyparrhenia - Beckeropsis grassland
- Swamp forest
- Seasonally flooded grassland
- Southern Guinea savanna

Derived from moist evergreen forest

5 4 3 2 1 0   10   20 Miles

5   0   10   20   30 Kilometres

Fig. 6. Vegetation Map of the Bamenda Highlands showing the Dom – Kejojang forest block. Escarpment denotes 1500m contour. Reproduced with permission from Hawkins & Brunt (1965)

38

The part of Hawkins & Brunt's (1965) map 10 (Fig. 6) that corresponds with Dom shows a patch of forest facing westwards on the edge of the high lava plateau. It is not subdivided further into vegetation types; in fact, throughout this map, forest is only shown as a single mapping unit. The accompanying text, however (Hawkins & Brunt 1965, 1:208-214) shows that the authors recognised two forest types, based on altitude and divided by the 5000ft. contour (=1500m). Above this contour they recognise 'moist montane forest' and below it 'moist evergreen forest' (lowland evergreen forest). However, they deliberate at length (p. 213) about the possible existence of a 'transition forest' (between these two types). This is prompted by their reading of Lebrun's work in 1935, on altitudinal classification of the forests of Kivu and Ruwenzori, in which a distinct forest type was recognised as occurring between 3300ft and 5250ft, this is c. 1000-1600m.

It is clear that their 'transition forest' equates to what is referred to below, here and by other more recent works in this part of Africa, as submontane forest. Hawkins and Brunt conclude with a statement that 'The few remnants of the original forest cover at lower altitudes in the Bamenda area, have not been sufficiently carefully studied to permit the suggestion that such a Transition Forest also occurred there.' Had they the opportunity to visit and study the forest at such sites as Dom and Bali Ngemba, they would have reached a very different conclusion. Instead, within this altitudinal range, they were only able to locate (p. 214) 'small stands of tall forest, protected for religious reasons, also remain...and near Bambui village....*Pterygota*.... 150-200ft high.' The significance of the *Pterygota* is discussed below.

Since Hawkins & Brunt's (1965) main geographical focus was what is today's NW Province of Cameroon, the larger part of their treatment of vegetation was devoted to grassland and savanna, since this dominates the province. They conclude that grassland and savanna in the Bamenda Highlands is derived, developing in areas which have been cleared of forest by man, the savanna species having migrated from lower altitudes. From their analysis it is clear that the forest of the type seen today at Bali Ngemba was formerly widespread and common in the Bamenda Highlands, occupying the 'low lava plateau' on which such large towns as Bali, Bamenda and Kumbo sit, and which today are among the most densely populated parts of the Cameroon. It was probably this range of forest, also, that once covered the adjoining highland areas, in West Province, of the Bamilike people, as densely settled as NW Province and even more denuded of the original forest.

Letouzey's 1985 vegetation map of Cameroon classifies the forest corresponding with Dom into two types:

- Submontane forest 800 to 1900-2000m alt. (Letouzey mapping unit 117)

- Montane forest 1900-2000m alt. and above (Letouzey mapping unit 108).

Surrounding the forest-inselberg block that corresponds with Dom, Letouzey shows five anthropic types of vegetation:

- Mapping unit 182 plant communities on rock (our survey included this as 3. Submontane inselberg grassland and rock faces).

- Mapping unit 113 surrounds most of the forest except to the west. This is characterised as *Sporobolus africanus* (Gramineae) montane pasture (1600-2800m) with gallery forest and has been characterised elsewhere in the Bamenda Highlands by Cheek *et al.* (2000). This grassland type is believed to result from repeated firing and cattle-tramping after forest clearance, and is generally species-poor and of low conservation value. Our survey did not include this vegetation type.

- Mapping unit 136 borders on the forest to the west. This is characterised as shrubby savanna (1200-1600m alt.) and results from forest clearance followed by agriculture. Our survey did not include this vegetation type.

- Mapping unit 128 is shown as a small patch to the south of the forest block. Leouzey characterises it as 'submontane grassland and farmbush, 1200-1800 (-2000)m, more or less grazed and inhabited'. This is also believed to be derived from forest, but occurs at lower altitudes than *Sporobolus africanus* (Gramineae) grassland. It equates with the 'derived savanna' discussed by Hawkins & Brunt (1965: see above). Our survey includes this vegetation type as 4. Derived savanna.

- Mapping unit 121 abuts the southern tip of the forest patch. Letouzey characterises it as domestic countryside without hedges or with dispersed enclosures, 1000-2000m. Our survey did not include this vegetation type.

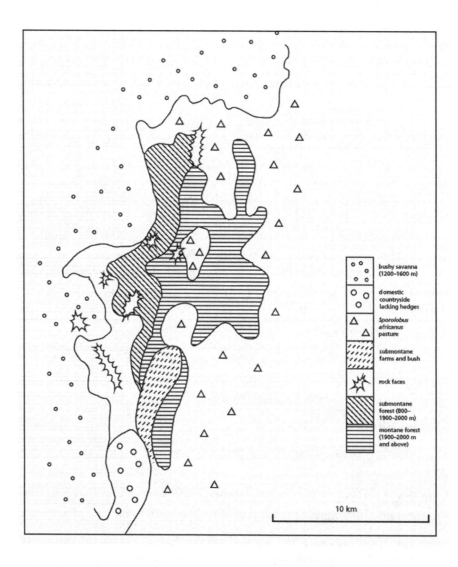

Fig. 7. Vegetation map of Dom and surrounding area, based on Letouzey (1985). Drawn by Martin Cheek.

CONVENTIONS USED IN VEGETATION TREATMENTS

In the following vegetation treatments list of species are given based on specimens and sight records made by a single team at a named locality so as to give a snapshot of the species assemblages in one area on one day. Species are listed in the order of collection and prefixed by data on the location and collectors. Species names in (parentheses) indicate additional taxa found in the same vegetation type but not noted at this location. (E) indicates an epiphyte.

Threatened taxa are those assessed as being threatened according to IUCN (2001) criteria and are included in more detail in our Red Data Chapter. They are indicated by being annotated as VU (vulnerable), EN (endangered), and CR (Critically Endangered) in the current chapter.

1. SUBMONTANE FOREST (1550 to 1900-2000m alt.)

The altitudinal definition of submontane forest used by Letouzey (1985: mapping unit 117) is adopted here, since it agrees with observations made at both Dom and Bali Ngemba, and also with other studies of altitude-defined forest types (Cheek *et al.* 2000, Cheek *et al.* 2004). The lower altitudinal limit of submontane forest in the works cited above is usually taken as the 800m contour. However, this altitude is not attained in the Dom area, which lies above the 'low lava plateau' (900-1200m alt.) which itself forms the perimeter of the Bamenda Highlands (Hawkins & Brunt 1965). The upper limit of submontane forest does occur within the Dom forest. As at Bali Ngemba, its demarcation with montane forest is less clear than that at e.g. Mt Oku and the Ijim Ridge (Cheek *et al.* 2000) in that several species that at Mt Oku and elsewhere in the Cameroon Highlands seem to characterise montane forest (1900-2000m and above) have only been recorded at Dom at 1500-1650m altitude. These taxa are:

Prunus africana (Rosaceae),1630m alt.
Pavetta hookeriana var. *hookeriana* (Rubiaceae), VU,1550m alt.
Nuxia congesta (Loganiaceae), 1640m alt.

This phenomenon needs further investigation.

At Bali Ngemba 17 taxa believed to characterise submontane forest as a whole at that location were listed. Only eight of these have been found so far at Dom. These are:

Salacia erecta (Celastraceae)
Garcinia smeathmannii (Guttiferae)
Ardisa staudtii (Myrsinaceae)
Chassalia laikomensis (Rubiaceae) CR
Cremaspora triflora subsp. *triflora*
    (Rubiaceae)

Psychotria psychotrioides (Rubiaceae)
Boehmeria macrophylla (Urticaceae)
Cyphostemma rubrosetosum
    (Vitaceae)

Notable absences at Dom are:

Deinbollia sp. 1 of Bali Ngemba (D. oreophila ined.) (Sapindaceae)

Memecylon dasyanthum (Melastomataceae) VU

Quassia sanguinea (Simaroubaceae) VU

Leptonychia sp.1 of Bali Ngemba (Sterculiaceae)

Synsepalum msolo (Sapotaceae)

Several of the absent species are endemics of the Cameroon Highlands. It may be that Bali Ngemba, in the S of the Bamenda Highlands, represents the northern extent of these species.

However, in compensation, the forests of Dom have a set of species, e.g. *Antidesma pachybotryum* (Euphorbiaceae), that do not occur at Bali Ngemba, or in forests of the Cameroon Highlands further to the South. *Antidesma pachybotryum* has its southern outpost in the Bamenda Highlands and its range extends further N and E through to the easterly running Adamoaua Highlands. Such species represent a set of submontane taxa that are perhaps better adapted to a lower rainfall, longer dry season or lower atmospheric humidity.

This pattern of species substitution, when compared with Bali Ngemba, occurs in most altitudinal bands in the forest at Dom, becoming less marked as altitude rises and species diversity declines.

1b) UPPER SUBMONTANE FOREST WITH *PTERYGOTA MILDBRAEDII* (1500-1700m)

Pterygota mildbraedii (Sterculiaceae) occurs only in the lower altitudes of submontane forest in the Bamenda Highlands. At Bali Ngemba it extends down to 1300m alt., but forest at Dom is only found at 1550-1600m alt., having been cleared for agriculture long ago at lower altitudes.

Pterygota mildbraedii (Sterculiaceae) is a deciduous canopy emergent, famous for making massive trees. Trees c. 60m tall and 4m diam. at ground level have been cited (Cheek. p.17 in Harvey *et al.* 2004). Being so distinctive and common in forest in this altitudinal band it is an excellent marker species for this vegetation type.

The canopy is closed, its height is c. 20m, the species evergreen. Common canopy species are *Newtonia camerunensis* (Leguminosae: Mimosoideae) CR, *Garcinia smeathmannii* (Guttiferae), *Cola anomala* (Sterculiaceae), *Uvariodendron fuscum* (Annonaceae), *Symphonia globulifera* (Guttiferae), *Strombosia scheffleri* (Olacaceae), *Chionanthus africana* (Oleaceae), *Bersama abyssinica* (Melianthaceae), *Ficus oreodryadum* (Moraceae), *Prunus africana* (Rosaceae).

The small tree-shrub layer extends to 5m tall, is dense and diverse with many rare species occurring in large numbers. The species lists below are based on a few hours observation and collection at one site close to the main Dom village.

Forest around Kinjinjang Rock, c. 1600m alt. 25th September 2006

N 60° 21' 13"; E 10° 35' 52", M. Cheek with Rene Ndifon, Fon of Dom, K. Tah, V. Nana and O. Sene.

Trees and Shrubs

Pterygota mildbraedii (Sterculiaceae)
Maesa rufescens (Myrsinaceae)
Trilepisium madagascariense
    (Moraceae)
Garcinia smeathmannii (Guttiferae)
Brucea antidysenterica
    (Simaroubaceae)
Antidesma venosum (Euphorbiaceae)
Uvariodendrum fuscum (Annonaceae)
    – flowering material needed
Leea guineensis (Leeaceae)
Psychotria peduncularis (Rubiaceae)
Psychotria schweinfurthii (Rubiaceae)
Oxyanthus okuensis (Rubiaceae) CR
Psychotria moseskemei (Rubiaceae)
    CR
Clausena anisata (Rubiaceae)
Dracaena laxissima (Dracaenaceae)
Eugenia gilgii (Myrtaceae) CR
Newtonia camerunensis
    (Leguminosae: Mimosoideae)
    CR
Allophylus ujori (Sapindaceae) EN
Deinbollia sp. 2 of Kupe
    (Sapindaceae)
Psychotria sp. A aff. *calva* (Rubiaceae)
Allophylus bullatus (Sapindaceae) VU

Ritchiea albersii (Capparaceae)
Bersama abyssinica (Melianthaceae)
Rothmannia hispida (Rubiaceae)
Chassalia laikomensis (Rubiaceae) CR
Ardisia dom (Myrsinaceae) CR
Strombosia sp. 1 of Bali Ngemba
    (Olacaceae)
Ficus oreodryadum (Moraceae)
Pavetta sp. A of Dom (Rubiaceae)
Cola anomala (Sterculiaceae)
Antidesma pachybotryum
    (Euphorbiaceae) VU
Raphia mambillensis (Palmae)
Symphonia globulifera (Guttiferae)
Psydrax sp 1 of Dom (Rubiaceae)
Prunus africana (Rosaceae)
Macaranga occidentalis
    (Euphorbiaceae) (pioneer)
(*Campylospermum flavum*)
    (Ochnaceae)
(*Strombosia scheffleri*) (Olacaceae)
(*Chionanthus africanus*) (Oleaceae)
(*Pauridiantha paucinevis*) (Rubiaceae)
(*Pavetta hookeriana* var. *hookeriana*)
    (Rubiaceae)
(*Entandrophragma angolense*)
    (Meliaceae) VU

Climbers

Cyphostemma rubrosetosum
    (Vitaceae)
Dioscorea schimperiana
    (Dioscoreaceae)
Rutidea sp. aff. *decorticata*
    (Rubiaceae)
Cremaspora triflora var. *triflora*
    (Rubiaceae)
Landolphia dulcis (Apocynaceae)
Mondia whitei (Asclepiadaceae)
Dioscorea preussii subsp. *preussii*
    (Dioscoreaceae)
Paullinia pinnata (Sapindaceae)
Dioscorea bulbifera (Dioscoreaceae)

Pararistolochia ceropegioides
    (Aristolochiaceae) VU
Adenia lobata (Apocynaceae)
Agelaea pentagyna (Connaraceae)
Clerodendrum silvanum var. *bucholzii*
    (Labiatae)
Raphiostylis beninensis (Icacinaceae)
(*Jasminum pauciflorum*) (Oleaceae)
(*Piper guineense*) (Piperaceae)
(*Gouania longipetala*) (Rhamnaceae)
(*Rubus pinnatus* var. *afrotropicus*)
    (Rosaceae)
(*Keetia venosa*) (Rubiaceae)
(*Sabicea tchapensis*) (Rubiaceae)
(*Clerodendrum violaceum*) (Labiatae)

Herbs, terrestrial & epiphytic (E)= epiphytic, hemiepiphyte or epilith

Impatiens hochstetteri subsp. *jacquesii*
    (Balsaminaceae)
Habenaria malacophylla var.
    malacophylla (Orchidaceae)
Aframomum sp. of Dom
    (Zingiberaceae)
Brillantaisia lamium (Acanthaceae)
Brillantaisia owariensis (Acanthaceae)
Asplenium aethiopicum (Aspleniaceae)
    (E)
Coleochloa domensis ined.
    (Cyperaceae) (E) CR
Tectaria fernandensis
    (Dryopteridaceae)
Elatostemma paivaeanum (Urticaceae)
    (E)
Asplenium preussii (Aspleniaceae)

Pilea rivularis (Urticaceae) (E)
Droguetia iners (Urticaceae) (E)
Chlorophytum comosum var.
    sparsiflorum (Anthericaceae)
Culcasia ekongoloi (Araceae) (E)
Aneilema beniniense (Commelinaceae)
Sanicula elata (Umbelliferae)
Amorphophallus sp. aff. *zenkeri*
    (Araceae)
Elatostema monticola (Urticaceae)
Selaginella sp. (Selaginellaceae)
Huperzia sp. aff. *dacrydioides*
    (Lycopodiaceae) (E)
Palisota mannii (Commelinaceae)
Anchomanes difformis (Araceae)
Cryptotaenia africana (Umbelliferae)
Setaria megaphylla (Gramineae)

Threats

This forest type faces immense threats from agriculture having been extensively cleared for this purpose. While some areas have been demarcated for protection (forest patches 1-3), we witnessed areas recently cut and burnt for agricultural crops in April 2005 and February 2006 (see photos in colour section). The understorey of some forest is planted with *Coffea arabica* (Rubiaceae). The forest at Kinjinjang Rock, which is documented above, is not contained within any of the three community protected forests, despite having the richest collection of rare species in the Dom area.

Forest edge with grassland

The transition between forest and grassland often harbours a distinct set of species. At Dom the transition needs more sampling since it was studied only at one site, and that briefly, where submontane forest with *Pterygota* bordered Derived Savanna:

Forest edge around Kinjinjang Rock c. 1630m alt. 25th September 2006

N 6°21' 17"; E 10° 35'56", M. Cheek with Rene Ndifon, Fon of Dom, K. Tah, V. Nana and O. Sene.

None of the species listed grows in open grassland, nor can survive in the shade of forest. About 100m length of transition was observed. The transition varied from abrupt to a band 10m wide with a woodland appearance.

Bridelia micrantha (Euphorbiaceae)
Dombeya ledermannii (Sterculiaceae)
    CR
Justicia striata subsp. *occidentalis*
    (Acanthaceae)
Croton macrostachyus
    (Euphorbiaceae)

Gnidia glauca (Thymelaeaceae)
Triumfetta cordifolia var. *tomentosa*
    (Tiliaceae)
Jasminum pauciflorum (Oleaceae)
Vitex doniana (Labiatae)
Diodia sarmentosa (Rubiaceae)

(*Shirakiopsis elliptica*)
(Euphorbiaceae)
Sesbania sesban (Leguminosae:
Papilionoideae)
Ensete gilletii (Musaceae)
Cyathea dregei (Cyatheaceae)

Phoenix reclinata (Palmae)
(*Margaritaria discoidea* var.
discoidea) (Euphorbiaceae)
Psorospermum aurantiacum
(Guttiferae) VU

Most of these species are very widespread, occurring throughout W Africa, or even tropical Africa, in this niche. However, both *Psorospermum aurantiacum* (Guttiferae) and *Dombeya ledermannii* (Sterculiaceae) are restricted to the Bamenda Highlands and areas nearby and hence has been assessed as threatened.

Submontane forest with *Pterygota mildbraedii* (Sterculiceae) (–1700m alt.) similarities and differences with Bali Ngemba:

Of the 14 species listed as characterising this vegetation type at Bali Ngemba eight occur at Dom:

Alangium chinense (Alangiaceae)
Kigelia africana (Bignoniaceae)
Carapa grandiflora (Meliaceae)
Strombosia schlefferi (Olacaceae)
Psychotria sp. A. aff. *calva*
(Rubiaceae)

Elatostemma paiveanum (Urticaceae)
Chlorophytum comosum var.
sparsiflorum (Anthericaceae)
Amorphophallus staudtii (Araceae)

As with species characterising the broader submontane forest belt described earlier, those absent from the *Pterygota mildbraedii* (Sterculiaceae) band include several Cameroon Highland endemics which may have reached their northern limit at Bali Ngemba: *Drypetes* sp. 3 of Bali Ngemba (Euphorbiaceae), *Psychotria* sp. A of Bali Ngemba (Rubiaceae) and *Tricalysia* sp. B aff. *ferorum* (Rubiaceae).

1c) EXTREME UPPER SUBMONTANE FOREST (1700 to 1900-2000m)

Characterised by the absence of *Pterygota* trees, this forest was documented at forest patch 1, also known as Kowi. The area studied was essentially a gallery forest, following a stream down-slope over a 100m alt. interval, the forest occupying a strip c. 50m wide on each side of the stream. *Pterygota mildbraedii* (Sterculiaceae) does occur in this forest patch, but below the altitudinal band that was studied.

Kowi Forest (Forest Patch 1 of Dom), c 1830-1930m alt., 27th September 2006

N 6° 21' 00"; E 10° 36' 29", M. Cheek with K. Tah, V. Nana and O. Sene. Guides Rene Ndifon and Kasimwo Wirsiy.

Canopy is 10-15 (-20)m high. The main tree species are evergreen: *Ficus lutea* (Moraceae), *Pouteria altissima* (Sapotaceae), *Symphonia globulifera* (Guttiferae), *Polyscias fulva* (Araliaceae), *Albizia gummifera* (Leguminosae-Mimosoideae), *Garcinia smeathmannii* (Guttiferae), *Clausena anisata* (Rutaceae), *Olea capensis* subsp. *macrocarpa* (Oleaceae), *Syzygium staudtii* (Myrtaceae), *Prunus africana* (Rosaceae), *Zanthoxylum leprieurii* (Rutaceae) and *Trilepisium madagascariense* (Moraceae).

The small tree-shrub layer is less dense than at lower altitudes, with ground cover c 5%.

Forest Patch 1 "Kowi" between c. 1830-2000m alt.

N 6° 21' 00", E 10° 36' 29" (taken at lowest alt. covered). This is gallery forest, canopy is 10-15 (-20m) high, ground cover c. 5 %.

Trees & Shrubs
Pouteria altissima (Sapotaceae)
Polyscias fulva (Sapindaceae)
Albizia gummifera (Leguminosae-
    Mimosoideae)
Garcinia smeathmannii (Guttiferae)
Olea capensis subsp. macrocarpa
    (Oleaceae)
Clausena anisata (Rutaceae)
Tabernaemontana sp. of Bali Ngemba
    (Apocynaceae)
Rhamnus prinoides (Rhamnaceae)
Pittosporum viridiflorum
    (Pittosporaceae)
Oxyanthus okuensis (Rubiaceae) CR
Chassalia laikomensis (Rubiaceae) CR
Ardisia dom (Myrsinaceae) CR
Deinbollia sp. 2 of Kupe
    (Sapindaceae)

Rytigynia sp. A of Kupe (Sapindaceae)
Dracaena fragrans (Dracaenaceae)
Symphonia globulifera var. suaveolens
    (Guttiferae)
Psychotria peduncularis (Rubiaceae)
Syzygium staudtii (Myrtaceae)
Prunus africana (Rosaceae)
Ficus lutea (Moraceae)
Alangium chinense (Alangiaceae)
Coffea liberica (Rubiaceae)
Solanum anguivi (Solanaceae)
Xymalos monospora (Monimiaceae)
Zanthoxylum leprieurii (Rutaceae)
Trilepisium madagascariense
    (Moraceae)
Carapa grandiflora (Meliaceae)
(Schefflera abyssinica) (Araliaceae)
(Garcina conrauana) (Guttiferae)

Herbs
Pteris togoensis (Pteridaceae)
Achyrospermum aethiopicum
    (Labiatae)
Hypoestes forskaolii (Acanthaceae)
Habenaria malacophylla var.
    malacophylla (Orchidaceae)

Epistemma cf. decurrens
    (Asclepiadaceae) (E) EN
Asplenium aethiopicum (Aspleniaceae)
    (E)
Panicum acrotrichum (Gramineae) VU

Climbers
Cissus aralioides (Vitaceae)
Clerodendrum sp. 1 of Dom (Labiatae)
Jasminum dichotomum (Oleaceae)
Psydrax kraussioides (Rubiaceae)
Leptoderris fasciculata (Leguminosae:
    Papilionoideae)
Mussaenda arcuata (Rubiaceae)
Rutidea sp. aff. decorticata
    (Rubiaceae)

Dalbergia lactea (Leguminosae:
    Papilionoideae)
Maytenus buchananii (Celastraceae)
Mikaniopsis paniculata (Compositae)
Adenia rumicifolia (Passifloraceae)
Adenia lobata (Passifloraceae)
Landolphia buchananii (Apocynaceae)

Threats

This forest appeared to be under no immediate threat. Its status as community forest, protected by the Dom people, with the help of the NGO ANCO appears to be crucial to its survival. Fires set in the adjoining grassland during the dry season pose a potential threat should any very dry years occur since the fire might then invade the forest. It is unknown as to whether cattle herds invade the forest or not as occurs elsewhere in the Bamenda Highlands causing deleterious effects to the forest understorey, but we saw no sign of this.

Similarities and differences with Bali Ngemba

Extraordinarily, all four of the commonest species in this forest type at Bali Ngemba were absent at Dom. These being *Xylopia africana* (Annonaceae) (VU), *Magnistipula butayei* subsp. *balingembaensis* (Chrysobalanaceae) (CR), *Drypetes* sp. aff. *leonensis* of Bali Ngemba (Euphorbiaceae) and *Beilschmiedia* sp. 1 of Bali Ngemba (Lauraceae). Moreover, of the 11 species restricted to this band at Bali Ngemba, only three, *Cuviera longiflora* (Rubiaceae), *Entandophragma angolense* (Meliaceae) and *Allophylus bullatus* (Sapindaceae) (VU) have been detected at Dom.

The *Xylopia* and *Beilschmiedia* are so common and distinctive at Bali Ngemba and easily recognised that if present at Dom it is unlikely that they would have been missed if they were present so it is likely that their natural range did not extend to the northern part of the Bamenda Highlands at all. It is imperative to search this altitudinal band of forest further at Dom and nearby since at Bali Ngemba and elsewhere in the Bamenda Highlands it includes some rare and threatened species: *Oncoba lophocarpa* (Flacourtiaceae) (VU), *Oncoba* sp. nov. of Bali Ngemba (Flacourtiaceae) and V*epris* sp. B of Bali Ngemba (Rutaceae). *Ternstroemia* sp. nov. (Theaceae) and *Dovyalis cameroonensis* (Flacourtiaceae) found in this altitudinal band further south, may also occur here. So far only a few hours of botanical team effort have been invested in exploring this forest type in the Dom area.

2. MONTANE FOREST, GRASSLAND AND SCRUB-FOREST EDGE
(1900-2000m to 2200m.alt.)

The extent of montane forest in the Dom - Kejojang area is hard to ascertain, but the 1:200,000 IGN map of the area (Nkambe) gives three spot heights exceeding 2000m to the E of Dom (see Fig. 3). The terrain is highly broken, so a visual assessment from the ground is impaired.

The highest altitude reached by our teams at Dom was 1930m alt. at the Kowi Forest (Patch 1), as measured by barometric and GPS altimeter, so we barely touched this vegetation type. A concerted effort is needed to redress this. Nonetheless, we recorded seven of the 11 species of canopy tree recorded at this altitudinal belt at Bali Ngemba (Cheek p. 21 in Harvey *et al.* 2004). These species were mainly found at much lower altitudes, as follows:

Canopy Trees
Agarista salicifolia (Ericaceae) 1630m
Bersama abyssinica (Melianthaceae)
    1610m
Ficus oreodryadum (Moraceae) 1630m
Clausena anisata (Rutaceae) 1600m

Syzygium staudtii (Myrtaceae) 1800m
Coffea liberica (Rubiaceae) 1830m
Gnidia glauca (Thymelaeaceae) (forest
    edge) 1630-1815m

Species of this belt at Bali Ngemba not seen at Dom were *Schefflera mannii* (Araliaceae) VU, *Cassipourea malosana* (Rhizophoraceae) and *Ixora foliosa* (Rubiaceae) (VU). Further species characteristic of this forest type at Mt Oku and the Ijim Ridge, where it is most extensively developed and studied, are *Arundinaria alpina* (Gramineae) and *Ilex mitis* (Aquifoliaceae), also not recorded by us from Dom. However, *Schefflera abyssinica* (Araliaceae), *Olea capensis* subsp. *macrocarpa* (Oleaceae), *Xymalos monospora* (Monimiaceae), *Rhamnus prinoides* (Rhamnaceae), *Rytigynia* sp. A of Kupe (*R. neglecta* of FWTA) (Rubiaceae) (understorey forest shrubs/tree), *Morella arborea* (Myricaceae), *Crassocephalum bauchiense* (Compositae) (both VU, rocky areas outside forest), *Impatiens sakerana* (Balsaminaceae) (VU) also characteristic of this forest at Oku and along the Cameroon Highlands from Mt. Cameroon and Bioko in the S, were recorded by us at Dom.

As at Bali Ngemba the curious phenomenon was found of what we have regarded as typically montane forest species (*Prunus africana* (Rosaceae), *Pavetta hookeriana* var. *hookeriana* (Rubiaceae) VU and *Nuxia congesta* (Buddlejaceae)) being recorded at Dom only at much lower altitudes (1550-1640m alt.). This phenomenon harmonises with the unexpectedly low altitudes at which the seven montane tree species listed above were found.

Threats to this forest at Dom community forest appear low from our limited observations, being as stated for extreme upper submontane-forest observed at Kowi Forest (patch 1) above. Elsewhere in the forest block in the Dom-Kejojang area, massive losses have occurred in recent years due to clearance for agriculture (see Baena, rear cover of this book).

3. SUBMONTANE INSELBERG GRASSLAND & ROCK FACES (1600-1930m alt.)

3a) ROCK FACES

Under this vegetation subtype are included a variety of rock outcrops seen in September 2006 on the path along the new Kumbo road from Kijume village to Forest Patch 1. These include small convex areas; flat, angled rock faces and cliffs. Their plant cover is always herbaceous, not woody, varying from scattered tussocks over bare rock; to a continuous carpet. Sampling of rock vegetation at Dom has been opportunistic and not systematic. It is extremely likely that many other rock outcrops occur, with species not yet catalogued for Dom. Further studies at Dom should specifically target rock outcrops.

The species assemblages indicated here should coincide with the 'saxicolous communities' mapped by Letouzey (1985) from aerial photography.

'Basalt pavement grassland' reported from Mt. Oku and the Ijim Ridge (Cheek *et al*. 2000:18-19) has not so far been found at Dom, neither have areas with peaty deposits, nor extensive seepages. This may account for the absence at Dom of several species occurring in this vegetation type at the Ijim Ridge, and also at Bali Ngemba (Cheek pp. 23-24 in Harvey *et al* 2004): *Drosera pilosa* (Droseraceae), *Utricularia pubescens* (Lentibulariaceae), *U. scandens* (Lentibulariaceae), *Xyris* (Xyridaceae), *Eriocaulon asteroides* (Eriocaulaceae) and *E. parvulum* (Eriocaulaceae).

Both these species of *Eriocaulon*, and the *Drosera pilosa* are Red Data species. Another rare species of Rock habitats in the Bamenda Highlands, *Bafutia tenuicaulis* var. *zapfackiana* (Compositae) recorded at Ijim, is also not yet known at Dom. So far no threatened species are known from this vegetation type at Dom. All the species documented are widespread in tropical Africa so far as is known.

The small overlap in species composition between this mapping unit at Dom with the same at Bali Ngemba is probably due to that being rocky grassland rather than the more or less bare rock outcrops studied at Dom. Closer overlaps occur with the cliff vegetation seen elsewhere in the Cameroon Highlands. For example *Gladiolus aequinoctialis* (Iridaceae) and *Coleochloa abyssinica* var. *abyssinica* (Cyperaceae) also occur on cliff faces at Bakossi (Cheek *et al* 2004: 43-48).

Rock outcrops

Four or five convex areas of probably basaltic rock 6-15m wide were surveyed, each with a differing species composition. When surveyed in September 2006 most species were in full flower, the rocks resembling a meadow garden.

N 6° 21' 36" E 10° 36'23", 1797m alt.

First rock outcrop, 10m across.

Drier parts with *Plectranthus bojeri* (Labiatae), *Kyllinga triceps* (Cyperaceae), *Drymaria cordata* (Caryophyllaceae), wet pockets with *Utricularia andongensis* (Lentibulariaceae).

Aeollanthus repens (Labiatae) 'moist rocky sites at high elevation' (FWTA)
Spermacoce sp. aff. *spermacocina* (Rubiaceae)
Bulbostylis densa var. *cameroonensis* (Cyperaceae) VU
Bidens barteri (Compositae)
Sporobolus festivus (Gramineae) 'crevices & rock outcrops' (FWTA)

Second rock outcrop, 20mx10m, 2m tall.
Scleria melanotricha var. *grata* (Cyperaceae) 'seasonally swampy places' (FWTA)
Pycreus atrorubidus (Cyperaceae)

Third rock outcrop, 20m high
Cyanotis barbata (Commelinaceae)

Fourth rock outcrop, low, with seepage:
Swertia mannii (Gentianaceae)
Antherotoma naudinii (Melastomataceae)
Trifolium baccarinii (Leguminosae: Papilionoideae)

Leaving the road and ascending to the saddle between two high points, further rock sites were found:

N 6°21'12", E 10°36'25", 1931m alt.
Rock c. 50m x 10m, flat, slope 45°, seepages.

Loudetia simplex (Gramineae) in drier parts
Tripogon major (Gramineae)
Sporobolus africanus (Gramineae)
Swertia mannii (Gentianaceae)
Utricularia andongensis (Lentibulariaceae)
Andropogon schirensis (Gramineae)
Scleria interrupta (Cyperaceae)

Bulbostylis densa var. *densa* (Cyperaceae)
Ascolepis protea subsp. *protea* (Cyperaceae)
Ctenium ledermannii (Gramineae)
Habenaria nigrescens (Orchidaceae)
Dissotis longisetosa (Melastomataceae) VU

Large vertical rock (cliff) facing NW (Collections by K. Tah & O. Sene)
Gladiolus aequinoctialis (Iridaceae)
Coleochloa abyssinica var. *abyssinica* (Cyperaceae)
Plectranthus tenuicaulis (Labiatae)
Lindernia abyssinica (Scrophulariaceae)

Crassula schimperi subsp. *schimperi* (Crassulaceae)
Selaginella thomensis (Selaginellaceae)

Vertical, mossy, shaded rock face
Utricularia striatula (Lentibulariaceae)
Trichopteryx elegantula (Gramineae)
Pilea tetraphylla (Urticaceae)

Clearly there is a great deal of variation in species assemblages, and in species, between each of the seven rock sites documented. This is due in part to ecological specialisation within the rock plant communities, some species being specific to damp shady rock faces, others preferring full sun, others being restricted to seepage areas on shallow slopes, while yet others prefer cliffs.

Since each additional site studied offered species not previously encountered, it would be sensible to survey more rock sites until the species accumulation curve flattens out.

Most of the species documented above are specialists of either montane rock or of thin soil habitats. Some species habitat notes from FWTA have been added to the species lists above to make this point. However, some of the species listed are only 'opportunistic' lithophytes. Dom seems to hold a rich and diverse community of rock species, if only partially known.

No threats are known for this vegetation type at Dom. Elsewhere in the Cameroon Highlands, e.g. in Bakossi, grazing, mining for road building material, and fires have been considered as threats (Cheek *et al.* 2004).

4. DERIVED SAVANNA (c. 1600m alt.)

The study site was grassland with a few trees at the foot of Kinjinjang Rock, an area largely surrounded by forest. Slopes were up to 45° from the horizontal. There was little direct evidence of post cultivation except for a few *Agave sisalana* (Agavaceae).

Bottom of slope, 1630m alt. 25th September 2006.

N 6° 21' 17", E 10° 35' 56", M. Cheek with Rene Ndifon, Fon of Dom, K. Tah, V. Nana and O. Sene.

Savanna trees made up <5% cover and were 3-8m high. The grassland area had c. 25% cover by shrubs or robust herbs 1m tall, but in places these increased to c. 75% cover and 1.5m tall. Grasses were mostly not yet in flower so are under-recorded.

Shrubs & Trees
Entada abyssinica (Leguminosae: Mimosoideae)
Protea madiensis subsp. *madiensis* (Proteaceae)
Combretum molle (Combretaceae)
Combretum fragrans (Combretaceae)
Terminalia mollis (Combretaceae)
Agave sisalana (Agavaceae) (planted)
(*Psorospermum densipunctatum*) (Guttiferae)
(*Syzygium guineense* var. *guineense*) (Myrtaceae)

Herbs
Spermacoce spermacocina (Rubiaceae)
Indigofera mimosoides var. *mimosoides* (Leguminosae: Papilionoideae)
Ocimum gratissimum subsp. *gratissimum* var. *gratissimum* (Labiatae)
Virectaria major var. *major* (Rubiaceae)
Dissotis brazzae (Melastomataceae)
Eriosema parviflorum subsp. *parviflorum* (Leguminosae: Papilionoideae)
Pavonia urens var. *urens* (Malvaceae)
Oldenlandia rosulata (Rubiaceae)
Trichopteryx elegantula (Gramineae)
Pseudarthria hookeri (Leguminosae: Papilionoideae)
Drimia altissima (Hyacinthaceae)
Pteridium aquilinum subsp. *aquilinum* (Dennstaedtiaceae)
Micromeria imbricata (Labiatae)
Aspilia africana subsp. *africana* (Compositae)
Nephrolepis undulata var. *undulata* (Oleandraceae)
Platostoma rotundifolium (Labiatae)
Dissotis perkinsiae (Melastomataceae)
Berkheya spekeana (Compositae)
Vernonia purpurea (Compositae)
V. smithiana (Compositae)
Crotalaria spp. (Leguminosae: Papilionoideae)
Leucas oligocephala (Labiatae)
Elephantopus mollis (Compositae)
Vigna gracilis var. *gracilis* (Leguminosae: Papilionoideae)
Alectra sessilifolia var. *senegalensis* (Scrophulariaceae)

Helichrysum forskahlii (Compositae)
Tephrosia vogelii (Leguminosae:
    Papilionoideae)
Lippia rugosa (Verbenaceae)
Dryopteris athamantica
    (Dryopteridaceae)

Kotschya strigosa (Leguminosae:
    Papilionoideae)
Swertia eminii (Gentianaceae)
Kyllinga triceps (Cyperaceae)
Indigofera arrecta (Leguminosae:
    Papilionoideae)

This vegetation is fairly diverse (not recently burnt or heavily grazed) and representative of the grasslands that dominate the Bamenda Highlands of NW Province in Cameroon today. These are thought to be largely secondary, that is, derived from clearance of the original forest that is now entirely absent apart from a few patches (Hawkins & Brunt 1965). Sampling of derived savanna remains incomplete at Dom. It was sampled incidentally at other sites en route for the forest patches. In some areas it verges on inselberg grassland where soils are shallow over rock, and in others montane forest scrub and transition where the vegetation is more densely shrubby, usually adjacent to forest areas. Such variants will have a higher species diversity and a higher content of restricted range species that can be expected to have value for conservation.

The diversity of savanna tree species at Dom is low, perhaps due to the high altitude. Elsewhere in the Bamenda Highlands *Annona chrysophylla* (Annonaceae), *Lophira lanceolata* (Ochnaceae), *Daniellia oliveri* (Leguminosae: Caesalpinioideae), *Hymenocardia acida* (Euphorbiaceae), *Terminalia glaucescens* (Combretaceae), *Nauclea latifolia* (Rubiaceae), and *Piliostigma thonningii* (Leguminosae: Caesalpinioideae) are found (Hawkins & Brunt 1965). However, savanna with mainly trees of *Entada abyssinica* (Leguminosae: Mimosoideae) is held to be representative of secondary vegetation on lava-rich soils of the plateau according to Hawkins & Brunt (1965).

The origin of this vegetation type in the context of submontane forest in the Bamenda Highlands is discussed in more detail in Cheek (pp. 24-26 in Harvey *et al* 2004).

Almost all of the species listed above are very widespread in W and C Africa, or even throughout tropical Africa. None is threatened. This vegetation type is not threatened by man, but rather is increasing slowly, as the last remnants of original Bamenda Highlands forest are steadily cleared, farmed, degraded, and finally left as fallow or rough grazing.

THREATS TO THE FORESTS OF DOM AND KEJOJANG

Martin Cheek

Herbarium, Royal Botanic Gardens, Kew, Richmond, Surrey, TW9 3AE, UK

Thanks to the united efforts of the people of Dom, lead by their Fon, guided by ANCO and supported by their DO at Nkor, the prospects for forest surviving at Dom look good. It has been recognised that the drying-up of streams, the water supplies of the six village quarters during the dry season, has been caused by converting the forest to farms. The cutting of forest has stopped and the people of Dom have formed patrols to safeguard it. The boundaries of the forests have been demarcated and mapped. To expand the forest, areas of grassland near the forest patches have been set aside for reforestation, native tree seedlings raised and planted by the community effort to create new forest. This progress deserves praise. The sponsorship of the venture by IUCN-NL at the request of ANCO, has been rewarded.

But Dom is just part of what, just 40 years ago, was a great forest, 'Kejojang' or, on maps 'Kejodsam'. Once joined to the forest of Oku to the south, the decline of 'Kejodsam' was documented by BirdLife International workers. Mackay (1994) stated that satellite imagery from 1987 showed 'Kejodsam' forest to be 30km², but only c. 5km² survived in 1994. Mackay believed then that it 'may have a long term future'. However a later BirdLife survey (Yana Njabo and Languy 2000) showed that only c. 5ha remained (a decline of 90% in forest area over 6 years), and that the outlook of this remnant was bleak. A road had been opened through the forest allowing access to farmers. Yet it seems that while a dramatic decline had occurred, far more than 5ha remained and remains today. It is probable that the overly negative estimate of 5ha was based on a ground survey only. Such is the unevenness of the terrain that forest patches in hollows or hidden by ridges, cannot be seen unless the area is combed on foot, a very difficult operation. As reported in this volume (Baena, rear cover), forest has indeed declined in the 'Kejodsam' area. However, not by 90% in 6 years from 1987, but by a still sad and depressing 50% by 2003. Not 5ha but c. 1242ha of forest remain. Apart from Dom, the largest surviving forest areas of Kejojang may be those of Mbinon village some kilometres to the North, where hundreds of hectares of forest still survive. There is every reason to hope that, if political difficulties can be resolved, Mbinon can also begin to protect and enlarge its forests.

THE EVOLUTION OF THIS CHECKLIST

Martin Cheek

Herbarium, Royal Botanic Gardens, Kew, Richmond, Surrey, TW9 3AE, UK

The idea for the subject of this checklist, the forest of Dom in Kejojang, came from Paul Mzeka of ANCO. At this time, in 2004/2005, the Royal Botanic Gardens, Kew and National Herbarium of Cameroon teams were seeking worthy targets for a future conservation project in Cameroon. This project was to be the Red Data Plants, Cameroon project supported by the Darwin Initiative of the British Government. The first botanical surveys, in early 2005 and 2006 were carried out before the official inception of the project while the most recent, in September 2006, took place directly after the workshop in Yaoundé that launched it. Various administrative delays meant that duplicates of the specimens made in these three surveys did not reach Kew for final identification until October 2007, carried personally by Dr. Jean-Michel Onana, head of the National Herbarium of Cameroon. Within the following months, most of the specimens were finally identified by Dr Onana at Kew and by other specialists at Kew, both staff and visitors. These individuals are usually credited as the first authors on the family accounts that follow. However, it was not until October 2008 that Dr. Muasya of the University of Cape Town, a noted Cyperacene specialist, visited Kew. It was he who identified what is currently the 'flagship' species of Dom: the only obligate epiphytic sedge known in Africa and so far, unique to the forest of Dom: *Coleochlea domensis*. Early in 2009 as work began in compiling this book, Paul Mzeka of ANCO kindly wrote and sent several important introductory chapters. The remainder are mostly written by Martin Cheek. Overall co-ordination of the taxonomic part of the book was performed by Yvette (Tivvy) Harvey. By early July 2009 the book was ready for sending out to the publishers.

THE HISTORY OF BOTANICAL EXPLORATION AT DOM

Martin Cheek

Herbarium, Royal Botanic Gardens, Kew, Richmond, Surrey, TW9 3AE, UK

No other botanical exploration of the forests of Dom is known to us, apart from our own. Other forests that we have surveyed and for which we have published conservation checklists (Cable & Cheek, 1998; Cheek *et al.* 2000; Harvey *et al.* 2004, Cheek *et al.* 2004) have all had previous visits from botanists, some numerous or extensive. In other cases, areas nearby, of likely relevance to our checklist areas have been surveyed. The existence of significant forest surviving in the Dom area has been known to scientists and national authorities since at least the 1960s when it was clearly shown on maps based on aerial photography. See, for example, the maps of Hawkins & Brunt (1965) in the chapter on Vegetation. Despite this we have not traced a single specimen from Dom, nor from sites nearby, such as Nkor or Mbinon. The explanation for this lack of attention is probably that Dom is remote from main lines of communcation, such as the Bamenda Highlands ring-road, and in the past the access road may not have been good. (Fig.3). From the roads, the forest itself cannot be seen – just the usual grassland and scrub that now dominates the whole of the Bamenda Highlands now that its forests have been removed.

It is entirely thanks to ANCO (Apiculture and Nature Conservation Organisation) of Bamenda that we learned of the existence of Dom when we enquired of them of forests that might be targets for surveys under our planned Darwin-supported Red List project for the plants of Cameroon. We are grateful to ANCO not just for suggesting that we visit Dom, but transporting us there and making arrangements with the local authorities for our visits.

The following text documenting our exploration of Dom is taken from our expedition reports. Three visits were made to Dom by us, the first in April 2005, the second in February 2006 and the most recent in September 2006. These texts are reproduced intact. Some species referred to in them do not appear in the checklist proper since identification at Kew have shown them to have different names to those used in the field reports. For example, in the first report "*Allophylus conrauana* (?)" is referred to. We now know this to be *Allophylus ujori,* a new species restricted to submontane forest of the Bamenda Highlands. Likewise "*Milicia excelsa*", a valued timber tree in the Dom area, is not that species, of the fig family, Moraceae but another, *Pouteria (Aningeria) altissima* of the Sapotaceae. The *Milicia* does not occur at such high altitudes usually in any case.

Specimens, photographic and sight records made at Dom to date, the foundation of this work and cited throughout are as under the following series:

April 2005	February 2006	September 2006
F. Nana	J.-M. Onana	M. Cheek
B. Pollard	Y. Harvey	Cheek sight record
M. Corcoran	Onana sight record	
Corcoran sight record	Corcoran sight record	
Pollard sight record	Dom photographic record	

FIRST VISIT

Report on the R.B.G., Kew [Kew]– Herbier National Camerounais expedition to Dom Forest, N.W.Province, Cameroon
April 27th – 30th 2005.

Benedict John Pollard

Acronyms used in the report:

ANCO — Association of Northwest Conservation Organisations
BP — Benedict Pollard (RBG, Kew)
CW — ChesoWalters (ANCO)
EK — Ernest Keming (ANCO)
HNC — National Herbarium of Cameroon
KT — Kenneth Tah (ANCO)
MC — Marcella Corcoran (RBG, Kew)
NF — Nana Felicité (HNC)
PM — Paul Mzeka (Director, ANCO)
TS — Tantoh Sale (local guide)
VI — Verina Ingram (Dutch Development Agency)
VN — Victor Nfon (local guide)
YA — National Herbarium of Cameroon

Wednesday April 27th 2005

07:00 BP, MC and NF meet up with VI at Ideal Park Hotel. Journey to ANCO office in Bamenda, meet up with PM, KT and WC, then depart for Kumbo in 2 Hi-lux vehicles (1 belonging to VI and the other ANCO).

12:00 Arrive Kumbo – take breakfast and then to ANCO nursery. Several employees working on manual weeding of seedlings which are being grown on in black plastic bags. Shelter provided overhead by palm fronds. Area of nursery approximately 400 m^2, with an estimated 30,000 plants in cultivation. Plants in cultivation include *Calliandra*, *Prunus africana*, *Jacaranda* and threatened species: *Newtonia camerunensis*, *Oncoba* sp. nov., *Dovyalis cameroonensis*. Details provided in table below, based on information received June 10th 2005 from Kenneth Tah.

Species / genus	numbers of seedlings
Prunus africana	10710
Ilex mitis	2750
Calliandra sp.	2030
Croton macrostachyus	550
Millettia conraui	2800
Albizia gummifera	1240
Jacaranda sp.	2100
Cassia sp.	820
Leucaena sp.	3465
Acacia sp.	2100
Pittosporum viridiflorum	540
Oncoba sp nov. *	757
Newtonia camerunensis *	280
Eugenia gilgii *	600

* these species are Red Data species threatened with extinction and so are being grown by ANCO as part of conservation efforts to try and introduce these species into suitable habitats within their known distributional range. These sites are likely to

offer levels of protection higher than those in which they are currently known to occur.

By the end of June 2005, *Oxyanthus okuensis* should be included in the list. *Kniphofia reflexum* and *Alchemilla fischeri* subsp. *camerunensis* seeds as well should be ready in July. They are both now in flower (June 10th 2005) (*fide* Kenneth Tah).

13:30   Return to Kumbo for supplies, shopping in market. NF takes lead on this.
15:00   Depart for Dom forest by shortcut road over hills, the end of which was only recently cut and graded.
16:30   Arrive Dom forest and settle into accommodation. VI has own small tent, MC and BP stay in large 4-person tent, NF has own bed and room in zinc-roofed house, CW, EK and KT stay in second house nearby, which has freshly built pit toilet and screened bucket-shower area. Separate room is used to store food for the trip. Tents are erected just before very heavy rain. Channels dug to prevent water inundating tented areas. Visit to the chief of the village and presentation of a bottle of whiskey and 2 bottles of wine. Food prepared with help of local cook and NF.

Thursday April 28th 2005

0800   Whole team departs for forest and introduction to collecting techniques along the Sangnere quarter on the road to the community forest.   Interesting species encountered include *Allophylus conrauana* (?) [*A. ujori*](Sapindaceae), *Eugenia gilgii* in full fruit. The day is spent moving slowly upwards from c. 1600m at the base of the community forest collecting any known Red Data species, unfamiliar species and as many tree species as possible. This latter focus will enable us to elucidate the typical structure of the forest and the canopy species composition.   Red Data species encountered include: *Dombeya ledermannii*, *Chassalia laikomensis* and *Newtonia camerunensis*.   Other   species   vouchered   include: *Brucea   antidysenterica*, *Landolphia* sp., *Ficus* spp., *Bridelia* sp., *Garcinia* sp.
1500   Team starts return to camp, c. 1 to 1.5 hours' walk. Start pressing specimens at about 1630, just before heavy rain returns.

Friday April 29th 2005

0800   Team splits into two groups, one destined for the top of the community forest ridge (team I) and another (team II) to an unbotanised patch over the other side of the ridge.

Team I: BP, KT, VI, VN. Path from Dom to Javelong forest [not located on our maps], stopping en route to collect on an inselberg and search for Red Data species. Despite being similar in type and altitude to other inselbergs known to harbour Red Data species, none is found. Notably absent: *Eriocaulon asteroides*, *E. parvulum*, *Drosera pilosa*. No specimens made. Next, team proceeds along path from Javelong forest to top of Dom forest. Species collected include: *Garcinia* sp., *Carapa* sp., *Milicia excelsa* [misnamed locally, this is *Pouteria altissima*] and *Crassocephalum bauchiense*. Descend into forest from top end, collecting *Rothmannia urcelliformis* (?) [*R. hispida*], *Pterygota mildbraedii* – the largest tree seen in the vicinity at c. 35m tall. Head back to base camp at about 1300 but get caught in extremely powerful thunderous hailstorm! The whole understorey covered in hailstones, very cold and windy.

Team II: MC, NF, CW, TS. See database for itinerary and collections made.

Both teams were caught in the hailstorm and head back to camp where pressing continues once the sky has cleared.

Saturday April 30ᵗʰ 2005

VI, BP, NF, MC return to Bamenda along the ringroad and then onto buses for Mt Kupe (NF, MC), Douala (BP). VI returns home to Bamenda.

APPENDIX

List of species noted by BP but not collected (site records) in order of viewing. Specimens were not made on account of the short duration of the trip and the lack of equipment necessary for a more in-depth inventory. Some of the species were noted from savanna-type vegetation surrounding the 'forest proper'

Bridelia micrantha	Euphorbiaceae
Leea guineensis	Leeaceae
Maesa lanceolata [*M. rufescens*]	Maesaceae [Myrsinaceae]
Ocimum gratissimum	Labiatae
Croton macrostachyus	Euphorbiaceae
Psychotria peduncularis	Rubiaceae
Dioscorea spp.	Dioscoreaceae
Psorospermum aurantiacum	Guttiferae
Brillantaisia owariensis	Acanthaceae
Erythrococca cf. *hispida*	Euphorbiaceae
Newtonia camerunensis	Leguminosae - Papilionoideae
Plectranthus glandulosus	Labiatae
Rytigynia neglecta [sp. A of Kupe]	Rubiaceae
Paullinia pinnata	Sapindaceae
Mussaenda erythrophylla	Rubiaceae
Albizia gummifera	Leguminosae – Mimosoideae
Rauvolfia vomitoria	Apocynaceae
Polyscias fulva	Araliaceae
Entandrophragma angolensis	Meliaceae
Piper capensis	Piperaceae
Harungana madagascariensis	Guttiferae
Macaranga occidentalis	Euphorbiaceae
Desmodium repandum	Leguminosae – Papilionoideae
Pteridium aquilinum	Fern [Dennstaedtiaceae]
Phoenix sp.	Palmae
Boehmeria sp.	Urticaceae
Canthium sp.	Rubiaceae
Agarista salicifolia	Ericaceae
Prunus africana	Rosaceae
Cyathea sp.	Fern [Cyatheaceae]
Rubus pinnatus	Rosaceae
Solanum torvum	Solanaceae
Marattia fraxinea	Fern [Marrattiaceae]
Annona senegalensis	Annonaceae
Xymalos monospora	Monimiaceae
Impatiens sakerana	Balsaminaceae
Vernonia spp.	Compositae
Syzygium staudtii	Myrtaceae
Leucas deflexa	Labiatae
Schefflera abyssinica	Araliaceae
Cuviera longiflora	Rubiaceae
Pittosporum viridiflorum	Pittosporaceae

SECOND VISIT

Report On The RBG Kew – Herbier National Camerounais Expedition To Dom
Forest, NW Province, Cameroon
Monday 13th February – Saturday 18th February 2006

Marcella Corcoran & Yvette Harvey

Participants:

Kew team	Marcella Corcoran
	Yvette Harvey
YA team	Jean-Michel Onana
	Madame Nana
ANCO	Kenneth Tah
	Paul Mzeka
	George Ngangong [Kangong]
SIRDEP	Tinyu Cyprian
Dom (cook)	Quinta Bousaw
Dom (guide)	She Frederick
	Kasimwo Wirsey [Wirsiy]
	Johnson Nchianda
Driver	Abraham Njume

Mon Feb. 13th, Nyasoso to Bamenda, driven by Abraham Njume, in the Presbytarian General Hospital, Nyasoso vehicle. Met Mr Paul Mzeka, Madame Nana and Jean-Michel Onana at the Pastorial Centre, Upstation, Bamenda.

Tues Feb. 14th, Met Paul Mzeka in his ANCO office. Gave PM 12 copies of the Bali Ngemba checklist, and copies of posters appertaining to the Bamenda Highlands. Discussed plans for Dom, including meeting Kenneth Tah who is to join us. Met Tinyu Cyprian of SIRDEP (Society for Initiatives on Rural Development and Environmental Protection) to discuss a possible future field trip to the forest of Ako (Mbembe forest), close to the Nigerian border. Left for Kumbo at 9.30am and visited the ANCO nursery there, with George Ngangong [Kangong]. Seedlings are currently thriving and soon to be distributed/planted. Shopped for provisions at Kumbo and then arrived in Dom just before dark – passing areas where forest had recently been decimated.

Weds Feb 15th, Visited the District Officer. He is happy to allow us to explore the Dom Forest and keen to receive our final report. He mentioned that the deforestation is having a devastating effect locally – especially in creating water shortages. Local landlords, demanding payments in vegetables, are driving tenants to encroach further into pristine forest.

Headed into the forest of Kowi with She Frederick. This small submontane forest patch included *Garcinia* and *Newtonia* as the dominant taxa, the former part of the undisturbed undergrowth, the latter as the main canopy tree. On the edge of the forest, *Dombeya, Eugenia, Gnidia, Clematis, Gouania longipedunculata, Crassocephalum, Carapa, Tabernaemontana* and *Albizia* were encountered. The forest is not easily accessible, slopes of >45° and steep valley edges made traversing difficult. Marcella Corcoran severely damaged a rib in a fall during the exploration. Vegetation up to the edge of the forest had recently been burnt.

Thurs. Feb. 16th, Reached Sousi forest after a 3-hour climb from base, with Kasimwo Wirsiy (forest guide). Edge of forest is Guineo-Sudanian, high-altitude savanna, including *Combretum* (common) and *Entada, Gnidia, Terminalia, Kalanchoe, Lobelia columnaris* and many *Aframomum*. The forest 'proper' had more *Albizzia* than *Newtonia* (only a couple seen). Although expected, *Piptostigma* and *Annona senegalensis* were not seen. Other taxa encountered included *Macaranga, Croton oligandrum* [probably *C. macrostachyus*], *Leea guineensis* and *Entada*. Group very disappointed with the quality of this forest.

Friday Feb. 17th, Introduced to the Forest Officer and the Fon's Representative. Headed to Nguindjume forest, close to the Sousi forest, with Johnson Nchianda, the forest guide. This undisturbed gallery forest, on a steep slope, >45° in places. Plenty of *Newtonia* and *Albizia* were encountered, in addition, *Tabernaemontana, Garcinia* and *Cola cordifolia* [unlikely in submontane forest], with a herb layer dominated by *Dracaena* and *Palisota*. The group returned to Bamenda and spent the night at the La Verna Spiritual Centre after bidding goodbye to Kenneth Tah.

Saturday Feb 18th, Returned to the Bamenda ANCO office for a final meeting with Paul Mzeka. Discussed Ako in more detail and another potential project area, Ngie. Arrived in Nyasoso, with Madame Nana, late afternoon, and immediately started the specimen drying.

THIRD VISIT

Botanical survey in N.W. Province

Martin Cheek

For logistical reasons, collections were made using the Schweinfurth method and later dried in Yaoundé, at the National Herbarium, where databasing of the collection records (field books) was also initiated. The visit to Dom from Bamenda was made via Bambalang in Ndop Plain, where an initial survey of the Royal Forest was made on 23rd September 2006, at the request of ANCO.

Arriving in Bamenda from Yaoundé were Martin Cheek (MC - RBG Kew [Kew]), Victor Nana (VN – Nat Herb, IRAD), Olivier Sene (OS – Student, Univ. Yaoundé I and Darwin technician). On meeting at the ANCO office to review the programme with the head of ANCO Paul Mzeka, the group set off in the ANCO hilux with ANCOs Kenneth Tah (KT) and driver Jude Zenebiun.

Survey of Dom Forest, Noni; Sun 24 – Thurs 28 Sept 2006

Dom village is in Noni subdivision (under Bui division) near its capital Nkor. Its forest came to light when the people of Dom appealed to the Bamenda Highlands Forest Project (since closed) for help in managing the forest, and achieving community status, some years ago. ANCO began taking up this challenge early in 2005 with NC-IUCN assistance. ANCO suggested to RBG Kew that it would be useful to survey the forest, and so two botanical survey visits were made, in April/May 2005 and Feb/March 2006, by combined RBG/HNC/ANCO teams.

The team arrived late in the afternoon, after meeting with the Divisional Officer (DO) at his house in Nkor. The DO expressed his delight that ANCO had returned after some months' absence. Reduction in the dry-season flow of water, which he attributed to forest clearance in Noni, had resulted in 300 women appealing to him for help. Apparently Noni 'landgivers' who were traditionally accepted to have the right, had given out tracts of forest to Nso farmers in the N, in return for a tithe on the resultant crops. The DO expressed the high value that he placed on forest conservation, not just for watershed protection, but also to conserve rare species that might be of use for future generations. It later transpired that he had been active in visiting the forest at Dom, and in chairing meetings of the Forest Management Council.

At Dom, accommodation was provided free by the village to ANCO. While this was being arranged, MC visited the Fon of Dom, before reconnoitring the route with two forest guards to the nearest forest site, only 30 minutes walk from the village. The Fon expressed his intention of joining the survey team the following day.

Mon 25th Sept 06. Forest and grassland near village, below Kinjinjang Rock, c.1600m alt. Cheek 13404-13519, c.115 numbers.

The first day was spent studying the plants in this area together with guide Rene Ndifon, and the Fon of Dom (until 1pm). This area is not one of the three forest patches recognised by the community at Dom. However, it proved to contain numerous Red Data species, and also species that are new to science and restricted to the Cameroon Highlands. Forest species:

Red Data Species:	IUCN conservation status
Allophylus conraui [*A. ujori*] (one plant seen)	EN
Eugenia gilgii (one plant seen)	CR
Chassalia laikomensis	CR
Psychotria moseskemei	CR
Oxyanthus okuensis	CR
Deinbollia sp. 2 [of Kupe]	Not yet assessed
Dombeya ledermannii	CR
Newtonia camerunensis	CR
Psychotria sp. A aff. *calva*	Not yet assessed

This forest was restricted to a stream valley which was surveyed by following a footpath (6°21.221'N; 10°35.862'E) that led to farms of *Coffea arabica*, here cultivated under local shade trees. Traversing a steep-sided valley, densely forested with *Newtonia, Cola anomala, Pterygota* and *Garcinia*, we reached a grassland patch (6°21.277'N; 10°35.940'E) extending gently up slope towards the Kinjinjang Rock.

Scattered savanna tree species were *Terminalia, Combretum* sp., *Entada abyssinica* (common), *Maesa rufescens, Vitex doniana.*

Forest edge species were: *Croton macrostachyus, Phoenix reclinata, Gnidia glauca,* sp. ?*Margaritaria, Bridelia* spp, *Ensete gilletii* (about 20 plants seen), *Dombeya ledermannii.*

Montane tree species located were: *Prunus africana, Agarista salicifola, Nuxia congesta.*

Grassland species (fertile) collected were: *Platostoma oppositifolia, Satureja pseudosimensis, Urginea* [*Drimia*] *altissima, Pteridium* (suggesting fire influence), *Nephrolepis, Dissotis* spp., *Leucas oligocephala, Vernonia* spp, *Echinops* sp., *Kotschya* (sterile), *Swertia* spp. Most grass species were sterile (wet season not ended).

A thunderstorm began threatening unusually early at c.11am but no rain fell until 4pm. Meanwhile at Nkor, a very heavy downpour occurred.

Recording Noni names and uses

Thanks to the Fon, a significant number of Noni names of plants were gathered, it having been agreed with him that these would be incorporated in our proposed conservation checklist for Dom together with basic local uses, so that this aspect of the traditions of the Noni people would not be lost. This discussion took place before going to the forest, when all members of the visiting team, with the guides, met with the Fon at his palace to explain the purpose of the mission and to seek his approval for it. On subsequent days, at the request of Eric Chia, subsequent local names had been recorded using our guides as source, were scrutinised, and often found wanting, where any names were recorded at all. These guides often gave us Ndamso names (a neighbouring tribe) e.g. Rene Ndifon, or had very little knowledge of any local names at all. Eric Chia, having detected dubious or Ndamso names, referred them to the Fon who gave the Noni names required by both him and Eric.

Habenaria maitlandii

This species has only been collected once before, in June 1931, at a site ("Nchian" – Achain?) to the NE of Kilum-Ijim and has never been seen again. It may be extinct, and was discussed as such at the Red Data workshop, Yaoundé the previous week. Since Dom lies 20-30km from the site, a watch was kept for *Habenaria* in the hope that *H. maitlandii* might be rediscovered. On our first day at Dom a population of an *Habenaria* was found in full flower, not far from the village. Over several days, three more sites for the species were found which grew naturally in forest edge habitat and also in disturbed sites. Sadly, photos and drawings of this material were identified at Kew by Dr Cribb not as *H. maitlandii*, but a widespread species, *H. malacophylla*. However, two other species of *Habenaria* were found on the second full field day, growing in grassland, both with withered flowers. One of these, at the saddle to 'Forest Patch 1' had the slender leaves of *H. maitlandii*, but examination of its flowers are needed to be sure of its identification. The specimens await permits in Yaoundé (Nov. 2006).

Tues 26th Sept 06. Scrub, grassland, forest edge and rock habitats from Dom to the saddle of forest patch 1, along the dry season motorable road to Kumbo. Cheek 13520-13603, c. 83 numbers. The second full day was spent along the route to forest patch 1, collecting all fertile plant species encountered, many of which were those of lower montane scrub including Red Data species (identifications to be confirmed):
Crassocephalum bauchiense VU
Pentas ledermannii VU

Rock outcrops

Five rock outcrops of various sizes, shapes and aspects were encountered, with distinctive assemblages of species still to be identified. At this season they were still wet, and in full flower. Provisional identifications of the more conspicuous taxa are:

Bidens cf. *monticola* [*B. barteri*]	*Cyanotis barbata*
Labiatae? Aeollanthus	*Antherotoma ?naudinii*
Utricularia ?scandens	*Cyperaceae* spp.
Swertia spp.	*Microcharis* sp.
Scleria spp.	

Normally the rocks encountered were soft, and easily scored with metal. Those rocks on the upslope leaving from the Kumbo road to the saddle were softest, being scored by wood, being white, perhaps furthest decomposed, basalt?

Collections concluded at the saddle ridge above forest patch 1 since at 3pm a violent storm broke out.

Weds 27th Sept 06. Forest Patch 1: 6°21.007'N, 10 °36.486'E, 1831m alt., Cheek 13604-666 = 63 numbers. We returned to the saddle ridge and descended to forest patch 1. This forest occupies the slopes on each side of a stream, extending 100m or more from the stream on both sides. The forest seen was of very good ecological and conservation quality, showing little sign of any artificial damage of any sort whatsoever, apart from the occasional traversing paths used by cattle herds. The following taxa were amongst those recorded with specimens:

Garcinia conrauana	*Bersama abyssinica*	*Xymalos*
Epistemma sp. nov.	*Pouteria altissima*	*Olea capensis*
Deinbollia sp. nov. 2 [of Kupe]	*Coffea liberica* var. nov.	*?Syzygium staudtii*
Tabernaemontana sp. nov.	*Carapa* cf. *procera*	*Garcinia smeathmannii*
Alangium chinense	*Rhamnus prinoides*	*Maytenus*
Chassalia laikomensis (common)	*Pterygota mildbraedii* (lower alts)	*Oxyanthus okuensis*

The *Epistemma* was especially notable as a probable new species so far unique to Dom. Half a dozen plants were found on a fallen tree limb c. 20cm diam. Rootstocks, aerial stems, leaves and one old fruit were recovered. Plants were transferred to Dom (Eric Chia's sterile *Cola*) and to Bamenda (Kenneth Tah) in order to establish and produce accessible flowers, so that these might be preserved and the species worked out.

A troop of baboons were observed on a grassy hill beyond the forest when departing.

Forest patch 2: 6°21.593'N, 10°36.384'E 1797m. alt. A brief visit was paid to part of this patch on the return journey to Dom. However dark clouds, followed by heavy rain, reduced our ability to see within the forest, and only a few numbers were made. *Chassalia laikomensis* appeared dominant in the understorey.

Beyond the forest, under a line of *Gnidia* trees, the Dom community had, with ANCO support, in June 2005, planted *Prunus* etc seedlings to reforest the area. Now c. 30cm tall they were almost lost amongst the grass. Kenneth Tah monitored the plantings for survivorship and progress before we returned to Dom.

The final specimen bundle for Schweinfurthing was prepared that evening.

Habitat conservation assessment-Dom Forests

The high number of Red Data species are tabulated above in the account for Mon. 25th Sept.. Many of these species had been revealed on earlier surveys, such as the Critically Endangered *Newtonia camerunensis* which has the world's largest known population in the forests of Dom.

Although, disappointingly, the extremely rare *Habenaria maitlandii* was not confirmed from this survey, there is still a good prospect that it will be found in future from Dom, given the proximity to the type location and the high quantity of natural habitat in the area.

Perhaps the most exciting discovery on this visit was that of a probable new and possibly narrowly endemic *Epistemma*. However, flowering material is needed to confirm this discovery and we hope that Eric Chia is successful in catching it in this state on his farm at Dom.

The sectors of the forest surveyed on this visit with ANCO showed a very high degree of ecological intactness. However, we aware that other sectors of the forest, not visited on this trip, have witnessed active clearance and firing earlier in the year (our survey team's report from Feb.-March 06). The Dom forests still have important conservation value, but if destruction is continued this will soon cease to be the case.

Thurs 28th Sept 06. Dom to Bamenda. After pressing a few specimens brought by Eric Chia, we departed Dom, paying a courtesy visit to the DO who had been stranded by the breakdown of his car. Our plan to visit notables in Nkor was thwarted by their absence from town.

Before **Mbinon** we stopped at the large market to search out a notable from Mbinon to discuss possible future work at their forest. The people being unaware of his presence in town, we set off for Mbinon to find out that, after all, he was indeed at the market town. Being needful of reaching Bamenda by nightfall, and it already being late in the day, we departed. Along the way we were much delayed by an amplification of the car electrical problems that had begun on departing Bamenda. These were correctly diagnosed at last, near Kumbo, but further halts necessitated by the damaged cables caused the guest group to take public transport for the last part of the journey. All were reunited at the ANCO office with Paul Mzeka. After a brief recap of the trip, Cheek dined with Verina Ingram of SNV and Jaap Van der Walde, viewing their herbarium specimens, before catching the overnight bus to Limbe. V. Nana and Sene, with the specimens, took the overnight Amour Mezam to Yaoundé.

PRELIMINARY ETHNOBOTANY OF DOM

Martin Cheek[1] and the Fon of Dom[2]
[1]Herbarium, Royal Botanic Gardens, Kew, Richmond, Surrey, TW9 3AE, UK
[2]Kifume, Dom, Noni Subdivision, Bamenda Highlands, Cameroon

Dom is in the Noni area, although many names used on a day to day basis for plants come from the language of the people to the south east, Ndamso [Lamnso]. The names used in the table below are all provided and were vouched for as Noni by the Fon of Dom on 25th September 2006 when he joined our study group. Being based only on a day or two of field work the list is highly incomplete. Uses are indicated where available. Detailed medicinal uses are witheld. Noni names were recorded in the course of collecting specimens of all species encountered. Specimens were identified at Kew. About 70% of species had no Noni name, including all the rarer species, such as *Allophylus ujori*. An exception was the epiphytic Cyperaceae, *Coleochloa domensis* (Njing Njing), so far known only from the forests of Dom. "Njing Njing" however, is also used for other (perhaps all other) species of the family, not this *Coleochloa* alone. This also happens with *Dissotis* where different species are given the same name (see table below). It is to be hoped that this very preliminary study of the Noni names and uses of plants in the Dom area is not the last and that such studies begin now, so that this traditional knowledge can be recorded before it is lost. Younger Noni's should be schooled in this knowledge so that it can be passed down to future generations and not lost as elders pass away.

NONI NAME	SCIENTIFIC NAME	FAMILY	USES	COLLECTOR	NO.
Bien	*Leea guineensis*	Leeaceae		sight record	
Diam	*Solanum aculeastrum*	Solanaceae		Cheek, M.	13600
Dijor	*Pouteria altissima*	Sapotaceae		Cheek, M.	13634
Doum Doum	*Clausena anisata*	Rutaceae		Cheek, M.	13428
Ed Deh	*Prunus africana*	Rosaceae		Cheek, M.	13485
Ee Wor	*Commelina africana* var. *africana*	Commelinaceae	Medicinal.	Cheek, M.	13531
Effileh	*Vitex doniana*	Labiatae	Edible fruits	Cheek, M.	13518
Fijeh-Fa	*Antidesma venosum*	Euphorbiaceae		Cheek, M.	13418
Fintambursi	*Impatiens hochstetteri* subsp. *jacquesii*	Balsaminaceae		Cheek, M.	13400
Fiuw	*Sporobolus africanus*	Gramineae		Cheek, M.	13579
Fui	*Combretum molle*	Combretaceae		Cheek, M.	13490
Gom Ten	*Ensete gilletii*	Musaceae		sight record	
Gow-Gow	*Mondia whitei*	Asclepiadaceae		Cheek, M.	13429
Iroko	*Garcinia conrauana*	Guttiferae	Timber tree.	Pollard B. J.	1398
Keh-Ngum	*Landolphia dulcis*	Apocynaceae	For fences	Cheek, M.	13419
Kehtonton	*Dissotis thollonii* var. *elliotii*	Melastomataceae		Cheek, M.	13407
	Dissotis brazzae	Melastomataceae		Cheek, M.	13426
Kengum	*Ficus lutea*	Moraceae		Cheek, M.	13635
Kimbin	*Pilea rivularis*	Urticaceae		Cheek, M.	13450
Ki Keih	*Raphia mambillensis*	Palmae		sight record	
Kin Fieh	*Droguetia iners*	Urticaceae		Cheek, M.	13451
Kola	*Cola anomala*	Sterculiaceae	Eaten	Cheek, M.	13468
Kon-The	*Adenia lobata*	Passifloraceae		Cheek, M.	13442

NONI NAME	SCIENTIFIC NAME	FAMILY	USES	COLLECTOR	NO.
Kunchien	*Cyphostemma rubrosetosum*	Vitaceae	Tubers like cassava, used medicinally.	Cheek, M.	13399
Lawe	*Monanthotaxis littoralis*	Annonaceae		Nana Felicite	140
Leh	*Aneilema beniniense*	Commelinaceae		Cheek, M.	13458
Lidi Chani	*Uvariodendron fuscum*	Annonaceae	Fruits eaten by chimpanzee; firewood.	Cheek, M.	13420
Lui Lui	*Ocimum lamiifolium*	Labiatae	Medicinal	Cheek, M.	13528
Mlah	*Mussaenda arcuata*	Rubiaceae	Fruits eaten.	Cheek, M.	13602
Munjiy	*Ocimum gratissimum* subsp. *gratissimum* var. *gratissimum*	Labiatae		Cheek, M.	13410
Nam-Meh	*Plectranthus tenuicaulis*	Labiatae		Cheek, M.	13598
Ndanso	*Indigofera mimosoides* var. *mimosoides*	Leguminosae-Papilionoideae		Cheek, M.	13408
Ndoh	*Ficus oreodryadum*	Moraceae		Cheek, M.	13465
Njing Njing	*Coleochloa domensis*	Cyperaceae		Cheek, M.	13438
Njing Njing	*Cyperus dilatatus*	Cyperaceae		Cheek, M.	13601
Nken	*Chlorophytum comosum* var. *sparsiflorum*	Anthericaceae		Cheek, M.	13452
Nkun	*Asplenium preussii*	Aspleniaceae		Cheek, M.	13448
Nkun	*Tectaria fernandensis*	Dryopteridaceae		Cheek, M.	13449
Nkung	*Tectaria fernandensis*	Dryopteridaceae		Cheek, M.	13446
	Commelina benghalensis var. *hirsuta*	Commelinaceae	Medicinal	Cheek, M.	13537
	Crassocephalum crepidioides	Compositae	Medicinal	Cheek, M.	13535
	Coccinia barteri	Cucurbitaceae	Medicinal	Cheek, M.	13536
	Dioscorea schimperiana	Dioscoreaceae	Not eaten. Tubers hurt mouth if eaten.	Cheek, M.	13415
	Dioscorea bulbifera	Dioscoreaceae	Not eaten	Cheek, M.	13435
	Polygonum nepalense	Polygonaceae	Medicinal	Cheek, M.	13532
	Polygonum senegalense forma *albotomentosum*	Polygonaceae	Protects against thunder and lightning.	Cheek, M.	13667
	Passiflora edulis	Passifloraceae	Edible fruits	Cheek, M.	13533
	Platostoma denticulatum	Labiatae	Used in cooking	Cheek, M.	13423
Npwessem	*Vernonia*	Compositae	Bitter leaf	Cheek, M.	13483
Nsaan-Sah	*Kyllinga triceps*	Cyperaceae	Medicinal	Cheek, M.	13514
Nsuleh	*Eragrostis tenuifolia*	Gramineae		Cheek, M.	13603
Ntooloo	*Landolphia buchananii*	Apocynaceae		Cheek, M.	13633
Ntu-Bun	*Gladiolus aequinoctialis*	Iridaceae		Cheek, M.	13584
Sah-Ton	*Loranthaceae* indet. of Dom	Loranthaceae		Cheek, M.	13599
Sem or Seb	*Maesa rufescens*	Myrsinaceae		Cheek, M.	13411
Shem Mon Fuhm	*Pterygota mildbraedii*	Sterculiaceae	Wood is sawn for timber.	Cheek, M.	13480
Teleh	*Garcinia smeathmannii*	Guttiferae		Cheek, M.	13416

ENDEMIC, NEAR ENDEMIC & NEW TAXA AT DOM

Martin Cheek

Herbarium, Royal Botanic Gardens, Kew, Richmond, Surrey, TW9 3AE, UK

Of the six possibly endemic or near-endemic taxa listed below, five are considered as strict endemics, i.e. known only from Dom, and one as a near endemic (known from Dom and at one other site).

Erythrococca sp. aff. *anomala* (Euphorbiaceae) (possibly new to science and endemic to Dom).
Ardisia dom (Myrsinaceae) (endemic to Dom, described in this book)
Oxyanthus okuensis (Rubiaceae) (Mt Oku and Dom).
Pavetta sp. A of Dom (Rubiaceae) (possibly new to science and endemic to Dom).
Allophylus ujori (Sapindaceae) (Bali Ngemba, Mambilla Plateau).
Coleochloa domensis (Cyperaceae) (new to science and endemic to Dom, in press, reproduced this book).

NEW TAXA

In additon to the new taxa contained in the list of endemics above, the following eight taxa present at Dom are new to science (either newly described or requiring formal description), but are not endemic or nearly endemic, being more widespread and known from at least two other locations in addition to Dom.

Tabernaemontana sp. of Bali Ngemba (Apocynaceae) (also at Tabenken and Fosimondi).
Psorospermum cf. *tenuifolium* (Guttiferae) (also Bali Ngemba and Bamenda).
Ardisia bamendae (Myrsinaceae) (Mwanenguba – Tabenken, described in this book).
Strombosia sp.1 of Bali Ngemba (Olacaceae) (Mt Kupe – Dom).
Psychotria sp. A aff. *calva* (Rubiaceae) (Mt Kupe – Dom).
Rutidea sp. aff. *decorticata* (Rubiaceae) (probable new species, endemic to Dom).
Rytigynia sp. A of Kupe (Rubiaceae) (Mwaneguba – Dom).
Deinbollia sp. 2 of Kupe (Sapindaceae) (Mt Kupe – Adamoua).

The taxa new to science described or due to be described, solely or partly from material derived from Dom, thus numbers 12.

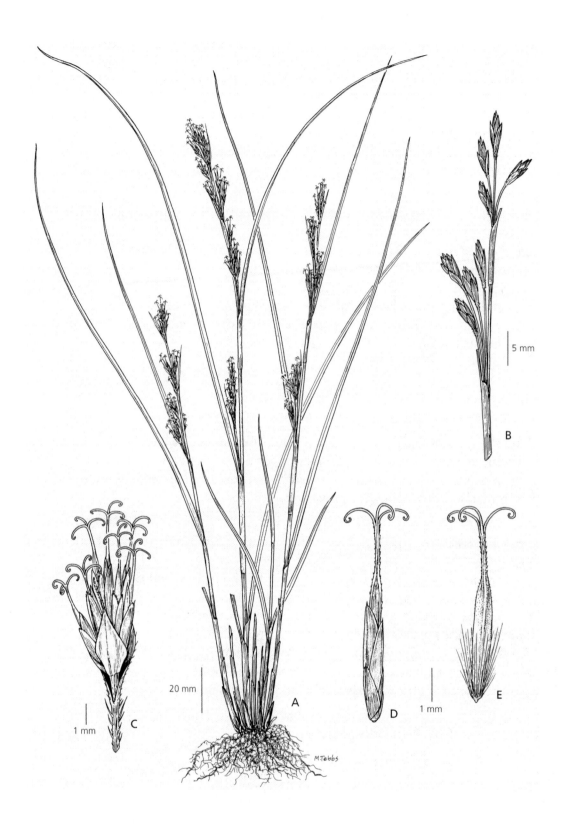

Fig. 8. *Coleochloa domensis*. **A.** habit; **B.** part of inflorescence; **C.** cluster of spikelets; **D.** spikelet; **E.** nutlet. All from the type. Drawn by Margaret Tebbs.

This paper is in press with the scientific journal, Kew Bulletin, publication is expected in 2010.

A NEW EPIPHYTIC SPECIES OF *COLEOCHLOA* (CYPERACEAE) FROM CAMEROON

A. Muthama Muasya[1], Yvette Harvey[2], Martin Cheek[2,] Kenneth Tah[3] and David A. Simpson[2]

[1]Department of Botany, University of Cape Town, Private Bag X3, Rondebosch 7701, South Africa.
[2]Herbarium, Royal Botanic Gardens, Kew, Richmond, Surrey, TW9 3AE, UK
[3]ANCO, P.O. Box 5150, Bamenda, NW Province, Cameroon.

Summary. *Coleochloa domensis* Muasya & D.A. Simpson, a new species of epiphytic sedge from submontane forest in Cameroon, is described and illustrated. Its conservation status is assessed as Critically Endangered (CR).

INTRODUCTION

The seven species of the genus *Coleochloa* (Trilepideae, Cyperaceae) are distributed in sub-Saharan Africa and Madagascar, where they grow on bare rocks and are characteristic of inselbergs (Govaerts *et al.* 2007; Goetghebeur 1998). The majority of the species are narrow endemics restricted to a single inselberg or country, with the exception of *C. abyssinica* and *C. setifera* which are more widespread occurring in several countries. Characters used to separate the taxa include leaf and nutlet size (Haines & Lye 1983).

Fieldwork in Dom forest in North West Province, Cameroon, in 2006 revealed an undescribed species of *Coleochloa*. This taxon was collected at two different forest localities separated by several kilometres, both times from fallen tree branches. The phorophyte (host tree) of the first collection (*Cheek* 13438) was not recorded. The second collection (*Cheek* 13647) however, was from a fallen *Ficus* (Moraceae) branch on the forest floor, where it was found together with various epiphytic orchids and the epiphytic *Epistemma decurrens* H. Huber (Apocynaceae-Asclepiadoideae). The branch had broken off from a height of 15-20m. While the branch had a sizeable population of the new species, no other material was growing in the understorey (Cheek pers. obs.). A follow-up visit to the site of the second collection in October 2008, with binoculars, revealed plants of the taxon growing high on other branches of original tree and a second colony about 100m away, growing on a branch of *Croton macrostachyus* Delile (Euphorbiaceae) at 7m above the ground (Tah pers. obs.). We describe this unique epiphyte here and compare it with the morphologically similar *C. abyssinica* (Table).

	C. domensis	C. abyssinica
stolon	absent	present
culm height (cm)	15–30	40–80
nutlet surface	pilose	smooth
style	persistent, tips recurved	breaking off, tips not recurved

Table. Comparison of morphology of *Coleochloa domensis* and *C. abyssinica*

Coleochloa domensis *Muasya & D.A. Simpson* **sp. nov.** *C. abyssinica* (A.Rich.) Gilly Typus: Cameroon, North West Province, Bamenda Highlands, Noni subdivision, near Dom village forest, around Kinjinjang Rock, 25 Sept. 2006 *Cheek* 13438 (holotypus K!; isotypus YA!).

A tufted perennial, lacking stolons or rhizomes. Culms 15–30cm tall, thickness at base including leaf sheaths 3–5mm. Leaves distichously arranged; sheaths open, margins overlapping, glabrous, 2–5mm long; blades flat, glabrous except upper surface of midrib which is densely hairy, 15.5–30.7cm long and 2–4mm wide, deciduous; ligule a dense band of white hairs, 1–1.5mm long. Inflorescence a stiff panicle (not pendulous), spikes 2-4 per leaf axil, peduncles unequal in length and longest to 3cm. Spikes comprising 6–10 spikelets; each spikelet bearing 4-6 florets; florets male or female. Spikelets to 3mm long with protruding seeds excluded. Glumes distichously arranged, to 2.5mm long, minutely papillose. Male flowers with 3 stamens, anthers to 1.2mm long. Nutlet brownish, pilose, 5-5.5mm long, including beak (to 3mm excluding beak), linear to lanceolate, trigonous in transverse section; style persistent on mature nutlets, base with tuft of white hairs, to 2mm long; stigmas 3, with recurved tips. Fig. 8.

DISTRIBUTION. Cameroon.

SPECIMENS EXAMINED. CAMEROON. North West Province, Bamenda Highlands, Noni subdivision, near Dom village, Forest around Kinjinjang Rock, 25 Sept. 2006, *Cheek* 13438 (holotype K!; isotype YA!); ibid. Oct. 2008, *Tah* s.n. (photo K!); Dom village, Forest patch I of Dom, 6°21'N, 10°36'E, 27 Sept. 2006 *Cheek* 13647 (K!; YA!).

HABITAT. Epiphytic in submontane tropical rain forest; 1600 m.

CONSERVATION STATUS. We assess this plant as Critically Endangered (**CR**) using the IUCN criteria (IUCN 2001) since it is only known from a single site which is under threat of forest clearance due to agricultural pressure. Overall it is estimated that as much as 96.5% of the original montane (including submontane) forest of the Bamenda Highlands has been lost (Cheek, 2004). GIS-based studies within another area of the Bamenda Highlands show that the loss of forest cover continues, with 25-30% lost between 1987-1995 (Moat, 2004). Surveys in other surviving submontane forest fragments of the Bamenda Highlands carried out by British and Cameroonian scientists in recent years have not yet revealed further sites for this species. The Apiculture and Nature Conservation Organisation (ANCO), a non-governmental forest conservation organisation based in Bamenda, has alerted the population of Dom, which manages the forest fragment inhabited by Coleochloa domensis, to the existence of this rare species and its conservation importance.

NOTES. *Coleochloa domensis* differs from *C. abyssinica*, the only congener occurring in West Africa, in being a smaller plant growing in tufts that lack any stolons. Nutlet characters are most conspicuously diagnostic, the nutlets being pilose and having persistent styles with characteristically recurved tips (Fig. 8). These characters are not observed in any other *Coleochloa* species.

This species appears to be one of the few sedges known to be obligately epiphytic, being the only one. *Coleochloa abyssinica,* while normally epilithic through most of its range, has been recorded as being epiphytic in Ethiopia (Lye & Pollard 2004). *Bulbostylis densa* (Wall.) Hand.-Mazz. var. *cameroonensis* Hooper has been observed growing epiphytically on one occasion (*Cheek* 7575 K!, YA!; Mt Kupe, Cameroon). The habitat of *C. domensis* is fairly similar to inselberg habitats occupied by most species in the genus. This region of the Cameroon Highlands is home to *Microdracoides* and *Afrotrilepis*, other members of the tribe Trilepideae, which are restricted to inselbergs.

ACKNOWLEDGEMENTS

We thank Margaret Tebbs for the illustrations. AMM acknowledges receipt of a Royal Society Incoming International Short Visit grant which funded travel to the U.K., during which the work describing this new species was undertaken. MC thanks the Darwin Initiative for support to the Red Data Plants project, Cameroon which funded the visit on which the species was discovered. ANCO, especially Paul Mzeka, are thanked for recommending that the forest of Dom be surveyed and for providing local logistical support for doing so. IRAD-National Herbarium of Cameroon is thanked for supporting fieldwork.

REFERENCES

Cheek, M. (2000). Quantification of Montane Forest loss in the Bamenda Highlands. In: M. Cheek, J.-M. Onana, & B. J. Pollard, *The Plants of Mount Oku and the Ijim Ridge, Cameroon, a Conservation Checklist*, pp. 49-50. Royal Botanic Gardens, Kew.

Goetghebeur, P. (1998). Cyperaceae. In: K. Kubitzki (ed.), *The Families and Genera of Vascular Plants* 4. Springer, Berlin.

Govaerts, R., Simpson, D.A., Bruhl, J., Egorova, T., Goetghebeur, P. & Wilson, K.L. (2007). *World Checklist of Cyperaceae*. Royal Botanic Gardens, Kew.

Haines, R.W. & Lye, K. (1983). *Sedges and Rushes of East Africa*. Eastern African Natural History Society, Nairobi.

IUCN (2001*). IUCN Red List Categories and Criteria: Version 3.1*. IUCN Species Survival Commission, Gland, Switzerland and Cambridge, U.K.

Lye, K. A. & Pollard, B. J. (2004). Cyperaceae. In: M. Cheek, B.J. Pollard,, I. Darbyshire, J.-M. Onana & Wild, C., *The Plants of Kupe, Mwanenguba and the Bakossi Mountains, Cameroon, a Conservation Checklist*, pp. 434-440. Royal Botanic Gardens, Kew.

Moat, J. (2000). Satellite Images of Mt Oku and the Ijim Ridge. In: M. Cheek, J.-M. Onana & B. J. Pollard, *The Plants of Mount Oku and the Ijim Ridge, Cameroon, a Conservation Checklist*. Royal Botanic Gardens, Kew.

ARDISIA DOM (MYRSINACEAE) A NEW SHRUB, UNIQUE TO THE FORESTS OF DOM

Martin Cheek

Herbarium, Royal Botanic Gardens, Kew, Richmond, Surrey, TW9 3AE, UK

Ardisia dom *Cheek* **sp. nov.** ab *A. bamendae* Cheek sepalis ad fructum appressis (non fructu reflexis), foliis glandulas nigras luce transmissa vel reflexa facile visibiles (non aegre visibiles) provisis, foliorum laminis 10–13(–15.8)cm longis (non 3.5–10cm) distincta. Typus: Cameroon, North West Province, Noni subdivision, Dom, Kowi Forest, 6°21'00" N, 10°36'29" E, 1831m alt., fr. 27 Sept. 2006, *Cheek* 13624 (holotypus K; isotypi BR, MO, P, WAG,YA).

Treelet, 1.5–2m tall, young growth with scurfy red indumentum. Main axis vertical, 0.5(–1)cm diam., terete, internodes c. 1cm long over the distal 15cm; lateral branches developed at each node, 25–60cm long, ascending at c. 45° from the vertical, basal 7–10mm dilated, laterally flattened, articulated with the main axis, extending longitudinally 10mm along the stem, 5mm wide, distal parts of lateral branch c. 2mm wide, flexuose, the leaves borne in one plane, leafless sides of branches with a longitudinal ridge; internodes 1–2cm long. Leaves elliptic or oblanceolate, 10–13(–15.8) × 3.4-5.3(–7)cm, apex acute or acuminate, acumen to 1 × 0.7cm, base cuneate and slightly asymmetric, margin subrevolute, serrate to dentate, teeth 2–3 per cm of margin, absent from the proximal, cuneate 2–3cm part of the blade, teeth triangular, projecting 1–1.5mm, sides each 2–3mm long, lateral nerves 8–10 on each side of the midrib, their basal half or more ascending towards the margin at c. 45° from the midrib, their upper portion, within about 5mm of the margin ascending more steeply, becoming parallel with the midrib and forming a submarginal nervelet with branches to the apex of each marginal tooth, lower surface and upper surface with black gland dots raised, 0.1–0.2mm diam., 850/cm^2, translucent brown dashes scarce, ratio c. 1:50 with black dots; lower surface midrib orange-brown, secondary to tertiary nerves, abruptly raised, yellow, contrasting conspicuously with the grey-black areolae, upper surface glabrous, lower surface glabrescent, when young red scurfy–arachnoid, large hairs stellate, 5–7 armed, 0.3–0.5mm wide, smaller hairs bushy, <0.1mm diam., 4–7-armed. Petiole canaliculate, 8–12mm long, 1.5–1.8mm wide. Infructescences absent from main axis, present only on lateral branches, absent from distal 3–5 nodes, usually U-shaped, 0.8–2cm long, 2mm wide with from 5–8 flower scars (the distal, youngest and shortest inflorescences) to 25 flower scars (the distal, oldest and longest inflorescences); pedicel scars exserted from the rachis, minutely puberulent. Bracts absent or inconspicuous. Pedicels straight, 10–15 × 0.5mm, dilating gradually from base to apex, minutely scattered puberulent. Calyx sepals 5, free, appressed to berry; sepals subtriangular, 1 × 1.2–1.4mm, with longitudinal black glands. Berry yellow, ellipsoid, 6–7 × 5–7mm, surface with raised black glands. Style persistent, 0.5mm long. Fig. 9.

DISTRIBUTION. Cameroon, North West Province, Noni subdivision, Dom community forests patches of the Kejojang Forest: endemic.

SPECIMENS EXAMINED. CAMEROON. North West Province, Noni subdivision, Dom, Kowi Forest, 6°21'00" N, 10°36'29" E, 1831m alt., fr. 27 Sept. 2006, *Cheek* 13624 (holotypus K!; isotypi BR!, MO!, P!, WAG!, YA!); Dom forest

Fig. 9. *Ardisia dom.* **A** habit, apex of main axis; **B** lateral branch; **C** leaf margin detail, adaxial surface, distal part of leaf; **D** leaf margin detail, abaxial surface, proximal part of leaf; **E** black leaf glands: detail in transmitted light, including a lateral nerve; **F** calyx, inner face showing black leaf glands; **G** fruit, lateral view, showing raised, black glands. **A-G** from *Cheek* 13624. All drawn by Andrew Brown. Scale bars: single = 1mm, graduated single = 5mm; double = 1cm; graduated double = 5cm.

around Kinjinjang Rock, 1600m alt., 6°21'13" N, 10°35'52" E, fr. 25 Sept. 2006, *Cheek* 13455 (K!, YA!).

ETYMOLOGY. Named (noun in apposition) for the forests of Dom and for my son of the same name.

HABITAT. Submontane evergreen forest with *Albizia gummifera, Pouteria altissima, Pittosporum viridiflorum, Olea capensis* subsp. *macrocarpa, Maytenus buchananii, Garcinia smeathmannii, Tabernaemontana* sp. of Bali Ngemba, *Clausena anisata, Oxyanthus okuensis, Chassalia laikomensis, Deinbollia* sp. 2 of Kupe, *Coleochloa domensis, Epistemma* cf. *decurrens*; 1600-1830m alt.

CONSERVATION STATUS. Since *Ardisia dom* is currently known from only two individuals, each in a separate patch of forest separated by several kilometres, and since these patches are part of the Kejojang Forest which has seen a reduction of c. 50% in the 15 years 1988-2003 (Baena, this book, rear cover), and since it is here estimated that the generation time of one plant may be 10 years, this species can be assessed as threatened using any of IUCN (2001) criteria A-D inclusive. However, here it is assessed as **CR D1** (fewer than 50 mature individuals known) since this recognises the highest level of threat for the species, that is, Critically Endangered. It is to be hoped that further searching at Dom will discover many more plants of *Ardisia dom* and that young plants can be raised from seed for incorporating in the forest restoration work currently being executed at Dom by ANCO (Apiculture and Nature Conservation Organisation) and the local populace.

Survey of other forest patches of the Kejojang Forest, such as near Mbinon, might also extend the range of *Ardisia dom.*

NOTES. *Ardisia dom* can be distinguished from *Ardisia bamendae* which is also likely to be found in the Dom area (especially at higher altitudes) by the features in the accompanying table. *Ardisia bamendae* is established in this volume, replacing *Afrardisia cymosa* sensu de Wit (1958), Hepper (1963) and *Ardisia kivuensis* Taton *sensu* Cheek et al. 2000. *Ardisia staudtii* is also known from Dom. As currently accepted (Taton 1979), *A. staudtii* is a widespread and somewhat variable species. The measurements indicated in the table are taken from the Dom material. Halliday (1984) unaccountably sank *A. kivensis* into the already very broadly circumscribed *A. staudtii*. This approach is rejected here.

	Ardisia dom	*Ardisia bamendae*	*Ardisia staudtii*
longest peduncle length	20mm	5mm	5mm
leaf shape	elliptic or oblanceolate	elliptic	oblanceolate
leaf size	10–13(–15.8) × 3.4-5.3(–7)cm	3.5–10 × 1.7–4.2cm	to 16 × 5.5cm
tooth length (leaf margins)	1–1.5mm	1mm	0.5mm
sepals in fruit	appressed	reflexed	not known
inflorescence bracts	absent or inconspicuous	c. 2mm long	c. 2mm long

Table of similarities and differences between *Ardisia dom, A. bamendae* and *A. staudtii.*

Ardisia dom is remarkable for the high density, large size and conspicuousness of its black leaf glands which ocupy 30-40% of the leaf area. The raised, pale yellow quaternary nervation of the lower surface contrasts distinctively with dark interstitial areolar areas. The very long, to 20mm, arched peduncles are unusual although they do also occur in *A. kivuensis sensu stricto*. The ellipsoid fruit and inconspicuous bracts are also distinctive.

Perennial peduncles

The longest peduncles of *Ardisia dom* occur in the most proximal nodes of the lateral branches. Moving towards the proximal nodes, peduncle length declines steadily, reaching only c. 5-8mm long 4-5 nodes below the stem apex (the distalmost 3-4 nodes are not reproductive). This suggests that peduncles can remain active over two or three seasons, their apices remaining dormant between reproductive periods.

ACKNOWLEDGEMENTS

Janis Shillito is thanked for typing the manuscript, Tim Utteridge for reviewing and earlier version of the manuscript, and Mark Coode for translating the Latin diagnosis.

REFERENCES

Cheek, M., Onana, J.-M., Pollard, B.J. (2000). The Plants of Mount Oku and the Ijim Ridge, Cameroon: A Conservation Checklist. Royal Botanic Gardens, Kew.

Halliday, P. (1984). Myrsinaceae. Flora of Tropical East Africa, Balkema, Rotterdam.

Hepper, F.N. (1963). Myrsinaceae. Flora of West Tropical Africa, Vol 2. Crown Agents, London

IUCN (2001). IUCN Red List Categories: version 3.1. Prepared by the IUCN Species Survival Commission. IUCN, Gland, Switzerland and Cambridge, UK.

Taton, A. (1979). Contribution à l'étude du genre *Ardisia* Sw. (Myrsinaceae) en Afrique tropicale. Bull. Jard. Bot. Nat. Belg. 49: 81-120.

De Wit, H.C.D. (1958) Revision of *Afrardisia* Mez (Myrsinaceae). Blumea, suppl. IV: 241-262.

ARDISIA BAMENDAE (MYRSINACEAE) A NEW NAME FOR THE MONTANE *ARDISIA* OF THE CAMEROON HIGHLANDS

Martin Cheek

Herbarium, Royal Botanic Gardens, Kew, Richmond, Surrey, TW9 3AE, UK

There is only one species of *Ardisia* that grows at altitudes of 2000m or more in Africa west of the Congo Basin. It is restricted to the surviving forest patches of the Cameroon Highlands between Mwanenguba and Tabenken. In the Flora of West Tropical Africa it is named as *Afrardisia cymosa* (Baker) Mez by Hepper (1963: 31) who followed de Wit's 1958 revision of the genus in treating *Afrardisia* as unique to Africa and distinct from otherwise pantropical *Ardisia*. Taton sank *Afrardisia* back into *Ardisia* and discovered that the type of *A. cymosa* Baker, from 600m alt in São Tomé, is distinct from the montane species of the Cameroon Highlands which he considered to be the same as that from Kivu. He chose a new name, *A. kivuensis* Taton for this taxon which we followed in our book on The Plants of Mt. Oku and the Ijum Ridge (Cheek *et al.* 2000: 149). However, the Kivu plant is quite distinct from the plants from the Cameroon Highlands as shown in the table below, so it is here formally named.

	Ardisia bamendae	*Ardisia kivuensis*
leaf-blade shape	elliptic	narrowly elliptic or oblong
blade length:breadth ratio	2:1 to 2.5:1	3:1 to 4:1
sepals in fruit	reflexed	appressed to fruit
pedicel indumentum	minutely puberulent	glabrous
bracts	2mm long	minute, inconspicuous

Table of similarities and differences between *Ardisia bamendae* and *A. kivuensis*

Ardisia bamendae *Cheek* ab *A. kivuensi* Taton foliorum laminis ellipticis longitudine quam latitudine 2–2.5-plo excedenti (non anguste ellipticis vel oblongis longitudine 3–4-plo latitudine excedenti), sepalis fructu reflexis (non ad fructum appressis) differt. Typus: Cameroon, North West Province, Mt Oku, Oku-Elak to the forest at transect KA, 2200m alt., fr. 27 Oct. 1996, *Cheek* 8440 (holotypus K, isotypi MO, P, WAG, YA)

Synonymy.   *Afrardisia cymosa* (Baker) Mez *sensu* de Wit (1958), *pro parte* (e.g. *Maitland* 1408).
                Ardisia kivuensis Taton (1979) *pro parte* (e.g. *Brunt* 604, 1210)
                Ardisia staudtii sensu Halliday (1984) *non* Gilg, *pro parte*.

SELECTED SPECIMENS. Cameroon, North West Region, Mt Tabenken, 10km SE Nkambe, 2200m alt., fr. 14 Nov. 1974, *Letouzey* 13526 (K!, P, YA); Oku evergreen forest, fl. 21 June 1962, *Brunt* 604 (K). Nigeria: North East State, Sardauna Prov., Kurmin Dodo, Chappel Waddi, 1650m alt., fr. 4 Feb. 1975, *J.D. Chapman* 3678 (FHI, K!).

Fourteen additional specimens cited as *A. kivuensis* are cited in Cheek *et al.* 2000 (149-150).

Ardisia bamendae is illustrated by de Wit (1958: 251, fig. 6) as *Afrardisia cymosa*.

HABITAT. Upper submontane and montane evergreen forests with *Schefflera abyssinica, S. mannii, Prunus africana, Bersama abyssinica, Syzygium staudtii, Nuxia congesta;* 1650-2620 m alt.

ETYMOLOGY. Named for the Bamenda Highlands in Cameroon which contains the principal surviving habitat for this species.

CONSERVATION STATUS. It has been calculated that 96.5% of the original forest of the Bamenda Highlands has been lost (Cheek in Cheek *et al.* 2000: 49). That which remains is also under great threat, for example, between 1988-2003 (15 years) about 50% of the Kejojang forest of which Dom, the subject of this book, is part, was lost (Baena, this book, rear cover). The losses of forest habitat of this species in that part of its range outside the Bamenda Highlands (e.g. the Bamboutous Mts) appear to have been even more extreme. Since these losses are ongoing and since the length of a generation of this species, while unknown, is certainly likely to be more than five years, the conservation status of *Ardisia bamendae* is here assessed as Endangered according to IUCN (2001), that is **EN A2b,c**.

LOCAL NAME AND USE. Echia (Oku language), children eat the sweet red fruit (*Cheek* 8440).

NOTES. Given that *Ardisia bamendae* is so distinct, it is remarkable that it has always been considered the same as the taxon east of the Congo basin, properly known as *A. kivuensis*. Within Cameroon the small, elliptic, thick leaves (so thick that the gland dots are very hard to see) with comparatively deep teeth, make the species easily recognised. In addition, no other species of *Ardisia* is known in Cameroon above 2000m alt. In the forest of Mt Oku and the Ijim ridge, this *Ardisia* is fairly frequent in the understorey (witness the large number of specimens, 14, cited in Cheek *et al.* 2000: 149-150), often in areas dominated by mass-flowering Acanthaceae. When at intervals of 7 or 9 years the Acanthaceae flower and die, the *Ardisia bamendae* shrubs are prominent on the otherwise bare forest floor.

ACKNOWLEDGEMENTS

Janis Shillito is thanked for typing, Mark Coode for the Latin translation, Tim Utteridge for comments on an earlier version of this text.

REFERENCES

Cheek, M., Onana, J.-M., Pollard, B.J. (2000). The Plants of Mount Oku and the Ijim Ridge, Cameroon: A Conservation Checklist. Royal Botanic Gardens, Kew.

Halliday, P. (1984). Myrsinaceae. Flora of Tropical East Africa, Balkema, Rotterdam.

Hepper, F.N. (1963). Myrsinaceae. Flora of West Tropical Africa, Vol 2. Crown Agents, London

IUCN (2001). IUCN Red List Categories: version 3.1. Prepared by the IUCN Species Survival Commission. IUCN, Gland, Switzerland and Cambridge, UK.

Taton, A. (1979). Contribution à l'étude du genre *Ardisia* Sw. (Myrsinaceae) en Afrique tropicale. Bull. Jard. Bot. Nat. Belg. 49: 81-120.

De Wit, H.C.D. (1958) Revision of *Afrardisia* Mez (Myrsinaceae). Blumea, suppl. IV: 241-262.

Fig. 10. *Lobelia columnaris* (Campanulaceae) by W.E. Trevithick

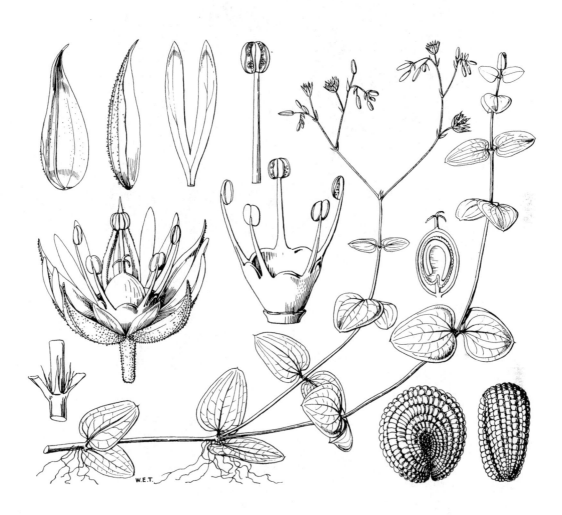

Fig. 11. *Drymaria cordata* (Caryophyllaceae) by W.E. Trevithick

Fig. 12. *Solanecio mannii* (Compositae) by W.E. Trevithick

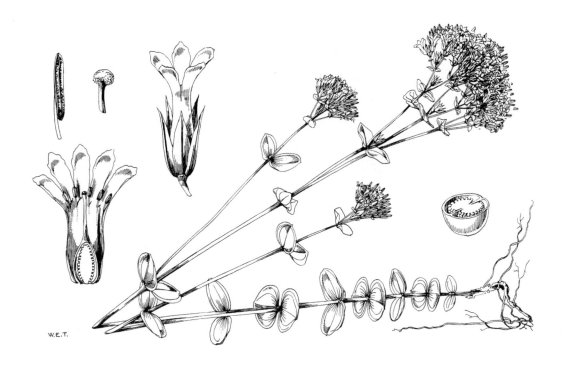

Fig. 13. *Sebaea brachyphylla* (Gentianaceae) by W.E. Trevithick

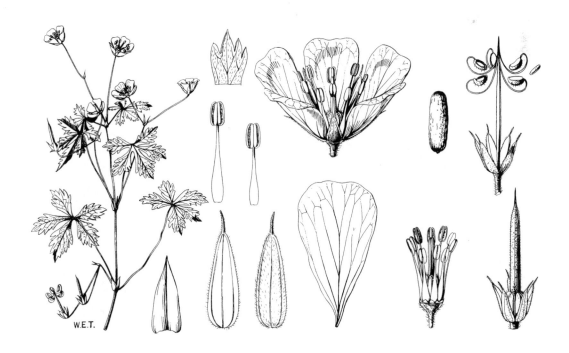

Fig. 14. *Geranium arabicum* subsp. *arabicum* (Geraniaceae) by W.E. Trevithick

Fig. 15. *Vitex doniana* (Labiatae) by W.E. Trevithick

Fig. 16. *Morella arborea* (Myricaceae) by W.E. Trevithick

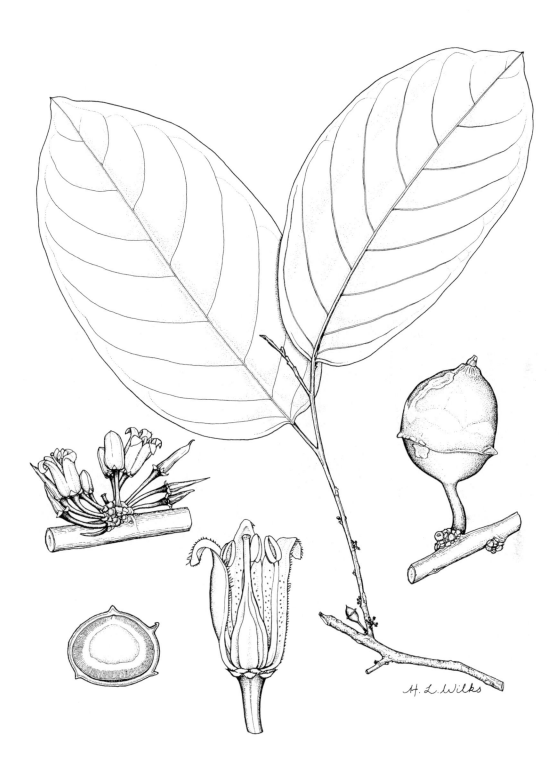

Fig. 17. *Strombosia* sp. 1 of Bali Ngemba (Olacaceae) by H.L. Wilks

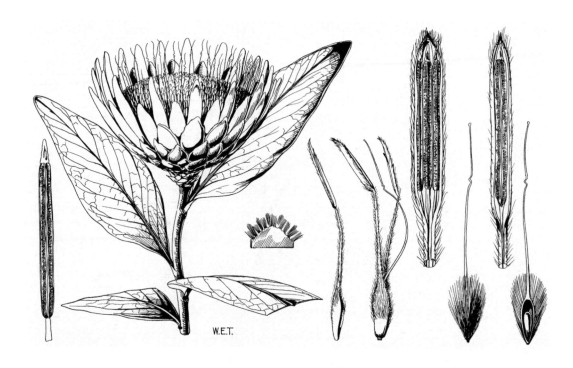

Fig. 18. *Protea madiensis* subsp. *madiensis* (Proteaceae) by W.E. Trevithick

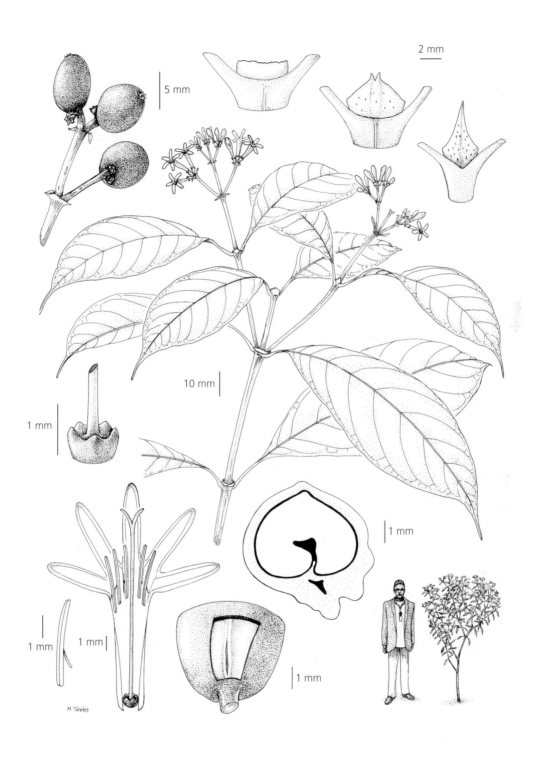

Fig. 19. *Chassalia laikomensis* (Rubiaceae) by M. Tebbs

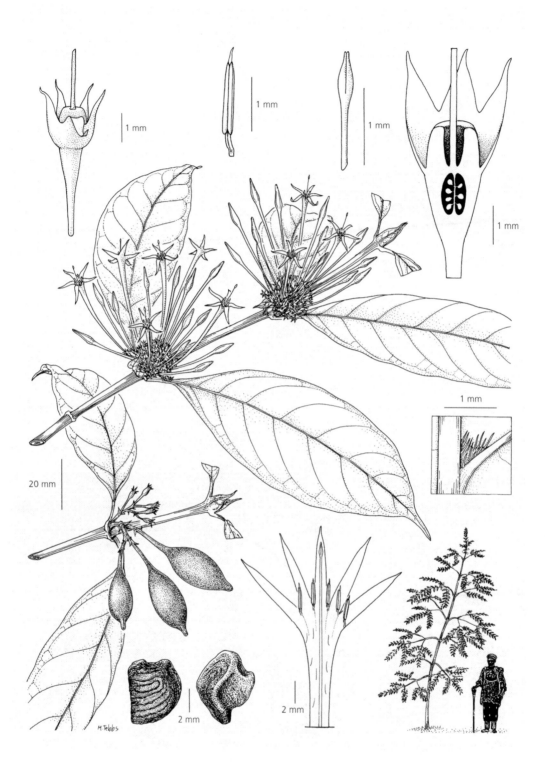

Fig. 20. *Oxyanthus okuensis* (Rubiaceae) by M. Tebbs

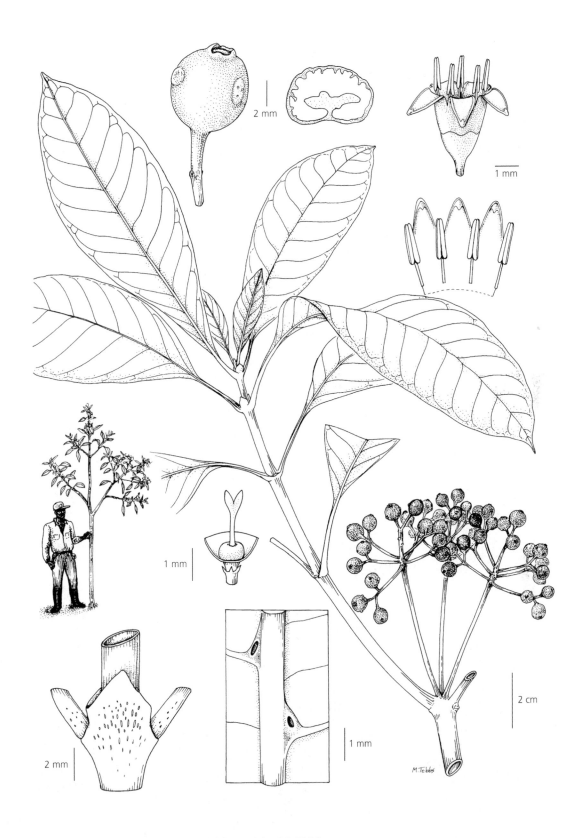

Fig. 21. *Psychotria moseskemei* (Rubiaceae) by M. Tebbs

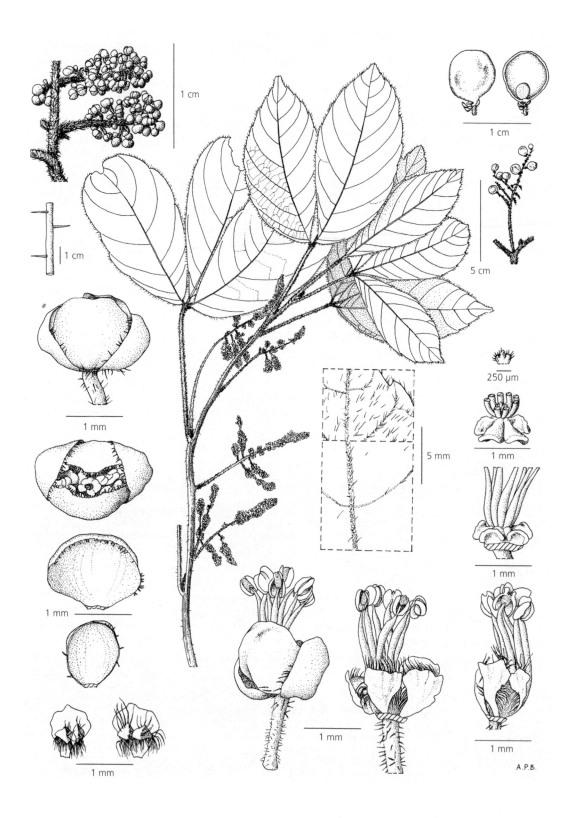

Fig. 22. *Allophylus ujori* (Sapindaceae) by A. Brown

Fig. 23. *Deinbollia* sp. 2 of Kupe (Sapindaceae) by A. Brown

Fig. 24. *Sporobolus africanus* (Gramineae) by W.E. Trevithick

A. Kowi Forest (dry season), February 2006, by Yvette Harvey
B. Kowi forest (wet season), September 2006, by Martin Cheek
C. Cattle grazing near Sousi forest, February 2006, by Yvette Harvey
D. Nguindjume forest edge with *Aframomum*, February 2006, by Yvette Harvey
E. Nguindjume forest watercourse, February 2006, by Yvette Harvey
F. Edge of Kowi forest, February 2006, by Yvette Harvey
G. Sousi forest from within, February 2006, by Yvette Harvey

Plate 1

A. Forest recently cleared for planting yams, February 2006, by Yvette Harvey
B. Deforestation within Sousi forest, February 2006 by Yvette Harvey
C. Burning to the edge of Kowi forest, February 2006, by Marcella Corcoran
D. Plant collecting in Kowi forest with (left to right) She Frederick, Yvette Harvey, Dr Onana, Mme Nana and Kenneth Tah, February 2006, by Marcella Corcoran
E. Kitiwum Nursery with Paul Mzeka (wearing orange), George Kangong, far right, and Mme Nana (fifth from left) and nursery staff, April 2005, by Marcella Corcoran
F. Kitiwum Nursery with Marcella Corcoran and George Kangong (chief nurseryman), February 2006, by Yvette Harvey

Plate 2

A. Fon of Dom, June 2009, by Kenneth Tah
B. Kenneth Tah, February 2006, by Yvette Harvey
C. Kasimwo Wirsiy (Community Forest Guard) with plant press on path to Sousi forest, February 2006, by Yvette Harvey
D. *Monanthotaxis littoralis* (Annonaceae), Onana 3612, February 2006, by Marcella Corcoran
E. *Rauvolfia vomitoria* (Apocynaceae), February 2006, by Yvette Harvey
F. *Tabernaemontana* sp. of Bali Ngemba (Apocynaceae), Onana 3629, February 2006, by Yvette Harvey
G. *Kigelia africana* (Bignoniaceae), Onana 3607, February 2006, by Yvette Harvey

Plate 3

A. *Lobelia columnaris* (Campanulaceae), February 2006, by Yvette Harvey
B. *Lobelia rubescens* (Campanulaceae), Harvey 308, February 2006, by Yvette Harvey
C. *Crassocephalum boughyeyanum* (Compositae), February 2006, by Yvette Harvey
D. *Launaea nana* (Compositae), February 2006, by Yvette Harvey
E. *Solanecio mannii* (Compositae), February 2006, by Marcella Corcoran
F. *Erythrococca membranacea* (Euphorbiaceae), Harvey 321, February 2006, by Yvette Harvey

Plate 4

A. *Plectranthus glandulosus* (flowering) (Labiatae), February 2006, by Yvette Harvey
B. *Stachys aculeolata* var. *aculeolata* (Labiatae), February 2006, by Yvette Harvey
C. *Newtonia camerunensis* (Leguminosae-Mimosoideae) seed on *Piper capense* (Piperaceae) leaf, Harvey 329, February 2006, by Yvette Harvey
D. *Globimetula oreophila* (Loranthaceae), Harvey 316, February 2006, by Yvette Harvey
E. *Clematis simensis* (Ranunculaceae), February 2006, by Marcella Corcoran
F. *Clematis villosa* subsp. *oliveri* (Ranunculaceae), Harvey 318, February 2006, by Marcella Corcoran

Plate 5

A. *Rubus pinnatus* var. *afrotropicus* (Rosaceae), February 2006, by Yvette Harvey
B. *Cuviera longiflora* (Rubiaceae), Harvey 320, February 2006, by Yvette Harvey
C. *Oxyanthus okuensis* (Rubiaceae) (fl), Onana 3602, February 2006, by Yvette Harvey
D. *Oxyanthus okuensis* (Rubiaceae) (fr), Onana 3602, February 2006, by Marcella Corcoran
E. *Psychotria psychotrioides* (Rubiaceae), Onana 3610, February 2006, by Yvette Harvey
F. *Sabicea tchapensis* (Rubiaceae), Onana 3614, February 2006, by Yvette Harvey

Plate 6

A. *Allophylus ujori* (Sapindaceae), Pollard 1380, April 2005, by Marcella Corcoran
B. *Brucea antidysenterica* (Simaroubaceae), Onana 3623, February 2006, by Yvette Harvey
C. *Dombeya ledermannii* (Sterculiaceae), February 2006, by Yvette Harvey
D. *Gnidia glauca* (Thymelaeaceae), Harvey 319, February 2006, by Yvette Harvey
E. *Amorphophallus staudtii* (Araceae), February 2006, by Yvette Harvey
F. *Anchomanes difformis* (Araceae), February 2006, by Yvette Harvey
G. *Coleochloa domensis* (Cyperaceae), July 2009, by Kenneth Tah

Plate 7

A. *Drimia altissima* (Hyacinthaceae), Harvey 317, February 2006, by Yvette Harvey
B. *Gladiolus dalenii* subsp. *andongensis* (Iridaceae), April 2005, by Marcella Corcoran
C. *Calyptrochilum christyanum* (Orchidaceae), Onana 3624, February 2006, by Yvette Harvey
D. *Phoenix reclinata* (Palmae), February 2006, by Yvette Harvey
E. *Aframomum* sp. of Dom (Zingiberaceae), February 2006, by Yvette Harvey
F. *Cyathea dregei* (Cyatheaceae), February 2006, by Yvette Harvey
G. *Pteridium aquilinum* subsp. *aquilinum* (Dennstaedtiaceae), with (left to right) Kenneth Tah, Dr Onana, Mme Nana, Kasimwo
   Wirsiy and Marcella Corcoran, February 2006, by Yvette Harvey
H. *Didymochlaena trunculata* (Dryopteridaceae), February 2006, by Yvette Harvey

Plate 8

RED DATA PLANT SPECIES OF DOM

Martin Cheek

Herbarium, Royal Botanic Gardens, Kew, Richmond, Surrey, TW9 3AE, UK

INTRODUCTION

As in the previous conservation checklists in this series (Harvey *et al.* 2004, Cheek *et al.* 2004), **all** flowering plant taxa recorded in the checklist have been assessed on a family by family basis for their level of threat, i.e. as threatened (CR – critically endangered; EN – endangered; VU – vulnerable), near threatened (NT) or of least concern (LC).

The main part of this chapter consists of taxon treatments, giving detailed information on the 22 Red Data species known to be present in the checklist area. Several of these taxa were assessed for the first time in the process of writing this book while others had been assessed as threatened in other publications and are reassessed here. All of the treatments have been commented upon by Craig Hilton-Taylor, the IUCN Red Data officer. IUCN rules do not allow acceptance of taxa unless they are either published or on the brink of publication. Consequently, the several new species to science that are not yet at this point, some of which are only known from Dom, have not been assessed and are therefore not mentioned in this chapter.

This, the introductory part of the chapter, details the methodology used in making the assessments, followed by a series of lists in which Red Data taxa are detailed by vegetation type.

ASSESSMENTS – METHODOLOGY

TAXA OF LEAST CONCERN (LC)

In the first place all taxa that were found to be fairly widespread e.g. Extent of Occurrence (IUCN 2001) greater than 20,000km^2 (and/or from 20 or more localities) were listed as LC. These facts were established principally using FWTA as an indication of range and number of collections sites, supplemented by other published sources, such as Flore du Cameroun, or by research into specimens at the Herbarium, Royal Botanic Gardens, Kew (K), in cases of doubt. Taxa which, by these measures, are widespread and common do not qualify as threatened under Criterion B of IUCN (2001), the main criterion used in our previous checklists for assessing threatened species. Under Criterion A, widespread and common taxa may, in contrast, still be assessed as threatened, but only if their habitat, or some other indicator of their population size has been, or is projected to be, reduced by at least 30% in the space of three generations, so long as this does not exceed 100 years.

THREATENED AND NEAR THREATENED TAXA

Those taxa in the checklist that were not assessed as LC, were then checked for level of threat using IUCN (2001). Most of these taxa were at least fairly narrowly endemic (restricted) in their distributions, e.g. endemic to Cameroon, or to NW Province, Cameroon and Nigeria. Criterion C, which demands a knowledge of the number of individuals of a species, was not used, since these data were not usually available. Criterion E, which depends on quantitative analysis to calculate the probability of extinction over time, was also not used. Criterion D was used for the two new species so far known only from Dom: *Coleochloa domensis* ined. and *Ardisia dom.*

i) Using Criterion B

About one-third of the assessments were made using Criterion B (usually B2ab), since the nature of the data available to us for our taxa lends itself to this criterion. Knowledge of the populations and distributions of most tropical plant species is mainly dependent on the existence of herbarium specimens. This is because there are so many taxa, most of which are poorly known and have never been illustrated. For this reason observations based only on sight-records are particularly unreliable and so undesirable in plant surveys of diverse tropical forest. Exceptions can be made when a family or genus specialist, working with a monograph at hand, or a proficient tree spotter identifying timber tree species. In contrast, surveys of birds and primates are not specimen dependent, since species diversity in these groups is comparatively low, and comprehensive, well-illustrated identification guides are available.

For the purpose of Criterion B we have almost always taken herbarium specimens to represent 'locations'. Deciding whether two specimens from one general area represent one or two locations is open to interpretation, unless they are from the same individual plant. Generally in the case of several specimens labelled as being from one town e.g. 'Bipinde', or one forest reserve e.g. 'Bafut Ngemba F.R.', these have been interpreted as one 'location'. Where a protected area has been divided into several geographical subunits, as at Mt Cameroon (see Cable & Cheek, 1998), and it is known that, say six specimens of a taxon occurring at Mt Cameroon fall into two such subunits, then this is treated as two locations. 'Area of occupancy' (AOO) and 'extent of occurrence' (EOO) have been measured by extrapolating from the number of locations at which a species is known. The grid cell size used for calculating area of occupancy is that currently advocated by IUCN, that is 4km². Information on declines in Criterion B has been obtained from personal observations, sometimes supplemented by local observers.

ii) Using Criterion A

The remaining assessments, about 60% of the total, were made using Criterion A. This criterion was used more than any other because montane and submontane taxa of the Cameroon Highlands that extend to the Bamenda Highlands were relatively easy to assess thanks to a Kew GIS unit studies of forest loss in two parts of the Bamenda Highlands, the first over 8 years in the Kilum-Ijim area (Moat in Cheek *et al.* 2000) and the second, over 15 years, in the Dom area (Baena, this volume).

These data have been used to estimate that during the last century, over 30% of the forest habitat of montane (above 2000m alt.) and submontane (800–2000m alt.) species has been lost in the Cameroon Highlands as a whole, so qualifying species with this range as Vulnerable. *Schefflera mannii* is an example of such a species. Formerly such taxa were often treated as unthreatened (e.g. Cable &

Cheek 1998) because they are secure on Mt Cameroon where little montane forest loss has occurred. Losses in this montane forest habitat have been highest in the Bamenda Highlands (and Bamboutos Mts) where c. 96.5% of original forest cover has been lost (Cheek in Cheek *et al.* 2000). Therefore species restricted to this range have been estimated as having lost over 80% of their habitat over the last century.

iii) Using Criterion D

Assessments using Criterion D generally depend on a knowledge of the global numbers of mature individuals (D1). This knowledge is generally not available for plant taxa in Cameroon except under VU D2. However, in the case of narrow endemics, where all known observations of a taxon have been made by us, the assessors, this criterion can be invoked.

CHANGES IN IUCN CRITERIA

It is to be hoped that there will be a continued moratorium on the changes in IUCN criteria which have been made in recent years. These changes made the work of assessors more difficult and have also reduced comparability of assessments made in different years. Assessments made in earlier Cameroonian checklists (e.g. Cheek *et al.* 2000) under IUCN (1994) criteria have been updated according to IUCN (2001) criteria where the taxa occur in the present checklist.

RED DATA SPECIES BY VEGETATION TYPE

The Red Data species of Dom are presented below under each vegetation type in which they occur. The vegetation types follow the classification used in the vegetation chapter. No species occurs in more than one vegetation type. More detailed information on the placement of the Red Data species within the subdivisions of each vegetation type is given in the vegetation chapter.

Species are listed below as they appear within the vegetation chapter, so that for example, within montane vegetation, forest species are listed first, and grassland-scrub-forest edge species last. In the Red Data taxon treatments that conclude this chapter the species are presented in the same order as in the main checklist, i.e. Dicotyledonae, Monocotyledonae, Pteropsida, then alphabetically by family and within family first by genus, then by species. The author(s) of the family assessment is given.

By far the most important vegetation type for threatened taxa is the submontane forest (17 Red Data taxa), with significant numbers also occurring in montane vegetation (3 Red Data taxa) and submontane inselberg grassland and rock faces (2 Red Data taxa). Derived Savanna has none.

1. SUBMONTANE FOREST (1550 to 1900–2000m alt.)

Pavetta hookeriana var. *hookeriana* VU
Chassalia laikomensis CR
Dombeya ledermannii CR
Allophylus ujori EN
Eugenia gilgii CR
Pararistolochia ceropegioides VU
Oxyanthus okuensis CR
Psychotria moseskemei CR
Ardisia dom CR

Antidesma pachybotryum VU
Coleochloa domensis ined. CR
Epistemma cf. *decurrens* EN
Panicum acrotrichum VU
Newtonia camerunensis CR
Entandrophragma angolense VU
Allophylus bullatus VU
Psorospermum aurantiacum VU

2. MONTANE FOREST, GRASSLAND AND SCRUB–FOREST EDGE (1900–2000 to 2200m alt.)

Impatiens sakerana VU
Morella arborea VU

Crassocephalum bauchiense VU

3. SUBMONTANE INSELBERG GRASSLAND & ROCK FACES (1300–1930m alt.)

Dissotis longisetosa VU

Bulbostylis densa var. *cameroonensis* VU

4. DERIVED SAVANNA (c. 1600m alt.)
No Red data taxa are known.

RED DATA TAXON TREATMENTS

The following Red Data taxon treatments are mostly modified and updated from accounts written earlier, both published (e.g. The Plants of Bali Ngemba, Harvey et al. 2004) and unpublished (e.g. Cameroon Plants Red Data book).

DICOTYLEDONAE

ARISTOLOCHIACEAE (M. Cheek)

Pararistolochia ceropegioides (S.Moore) Hutch. & Dalziel
VU A2c+3c
Range: Cameroon (6 pre-1978 coll.), Mt Kupe (19 coll., 3 sites), Dom (1 coll.); Gabon (3 pre-1978 coll.).
This canopy climber is easily spotted when fertile since the orange, bijoux flowers are produced in clusters at eye-level from the figure-of-eight shaped stems. It is widespread but fairly rare in semi-deciduous and submontane forest in the forest belt of Cameroon and northern Gabon. In Bakossi, although common (19 coll.), that part of its population occurring below 1000m alt. is under threat of forest clearance, while

there has been extensive forest loss at some of its known locations elsewhere, e.g. at Nlonako and Yaoundé.

Its presence at Dom is a range extension, so collection of flowering material is needed to confirm the identification.

Overall, 30% habitat loss for this species over the past three generations seems conservative given that it is apparently a long-lived perennial with generation times possibly exceeding 10-20 years. Forest in the Dom area declined by c. 50% in 15 years recently and is ongoing. However, habitat in other parts of the range, such as Gabon, tends to be much less threatened. Data on collections cited under range for this species is taken from Poncy (1978) (Adansonia 17 (4): 465–494).

The assessment above is updated and modified from that in Cheek *et al.* (2004).

Threats: forest clearance for timber and agriculture.

Habitat: semi-deciduous and submontane forest; 600–1600m alt.

Management suggestions: the Dom forests should be surveyed further to assess the abundance and confirm the identification of this species.

ASCLEPIADACEAE (APOCYNACEAE-ASCLEPIOIDEAE) (M. Cheek)

Epistemma decurrens H. Huber
EN B2ab(iii)
Range: Cameroon (Gotel Mts, slopes of Tchabel Ouade and Bali Ngemba F.R., probably Oku and Dom)

An epiphyte, known with certainty from only two gatherings (*Letouzey* 8606 and *Darbyshire* 378), its leaves resemble those of *E. rupestre*, while its flowers are 1cm wide, yellow. Here it is assessed as endangered since only two sites are known and at one of which, Bali Ngemba, it is acutely threatened.

Habitat: montane forest 2000m alt., flowering in June.

Threats: its forest habitat (e.g. at Bali Ngemba pers. obs. Cheek 2004) is under threat due to felling for timber, agricultural land and fires set by pastoralists. Most montane forest in the Cameroon Highlands has been lost.

Management suggestions: an attempt should be made to refind this species at its known localities and to protect it, if it survives. Further survey work in the Bamenda Highlands may reveal more sites. Several fruiting collections (made in Nov.) of *Epistemma* from near Lake Oku are known (*Pollard* 41, *Cheek* 8767) and from Dom in Sept. 2006 (*Cheek* 13648) which probably represent this species or possibly another, undescribed species of *Epistemma* that is equally threatened. These need to be re-collected when in flower (probably June) to secure a specific identification.

BALSAMINACEAE. (M. Cheek)

Impatiens sakerana Hook.f.
VU A2c
Range: Bioko; Cameroon: Mt Cameroon (6 coll.), Mwanenguba (2 coll.), Fosimondi, Bamboutos Mts (1 coll.), Bamenda Highlands (1 coll.), Bali Ngemba F.R. (3 coll.), Mt Oku (12 coll.), Dom (1 record).

This often robust, presumed perennial, locally common terrestrial herb has the highest altitudinal range of all Cameroonian *Impatiens*. Secure on Mt Cameroon, and probably Bioko, forest at Mwanenguba, the Bamenda Highlands and Bamboutos Mts has been under pressure from grassland fires set by graziers and for clearance for agriculture; forest in the Kilum-Ijim area outside the protected area having seen a reduction in cover of c. 25% in eight years of the 1980s and 1990s (Moat in Cheek *et al.* 2000). In the Dom area 50% forest loss has occurred 1988-2003 (Baena, this volume). Assuming a generation time of ten years, it is estimated that about 30% habitat loss may have occurred over its whole range in the last 30 years.

The assessment above is updated from that in Cheek *et al.* (2004) and Cheek in Harvey *et al.* (2004).

Habitat: understorey of montane forest; 2000-3000m alt.

Threats: see above.

Management suggestions: enforcement of protected area boundaries. Demographic studies are needed to elucidate generation time and ecological requirements of this taxon.

COMPOSITAE (M. Cheek)

Crassocephalum bauchiense (Hutch.) Milne-Redh.

VU B2ab(iii)

Range: Nigeria: Bauchi Plateau, Jos Plateau, Pawpaw Mt, Amedzefe (1 coll. each); Bioko (3 coll.); Cameroon: Bakossi (3 coll. at 2 sites), Mt Cameroon (11 coll.), Mt Oku (6 coll.), Dom (1 record).

This erect blue-flowered herb is known from nine montane sites in lower Guinea. If new sites are discovered it may be downgraded to NT, especially in view of the fact that few data are available regarding loss of its habitat.

The assessment above is updated from that in Cheek *et al.* (2004).

Habitat: open woodland, savanna, forest edge; 900–2000m alt.

Threats: clearance of trees for agriculture and wood.

Management suggestions: the effect of forest loss on this, and other rare, sun-loving *Compositae*, many of which share the same range of habitats, is poorly understood. Research is needed to redress this. Limited 'habitat destruction' at intervals of years, in the form of fires or limited long-fallow small-holder agriculture or small scale tree felling may be helpful to the survival of such taxa by keeping their habitat open and allowing establishment of probably short-lived perennials such as these.

Postscript: range of this taxon now believed to extend to Congo (Kinshasa) and Uganda (Beentje, *pers. comm.*) which may diminish the threat to this taxon when reassessed.

EUPHORBIACEAE (M. Cheek)

Antidesma pachybotryum Pax & K. Hoffm.

VU B2a,b(iii)

Range: CAR (1 collection, Bouar); Cameroon: (Dschang (2 collections), Bamenda, Mbouda, Dom, Tibati & Tibati-Mbanti (1 collection each).

Known historically from 7 sites in an arc extending from S to E along the Bamenda and Adamoua Highlands. Here assessed as vulnerable due to threats (see below), low number of both sites and low AOO (area of occupancy): the species appears not to be gregarious – no more than a single individual being noted at a site. It has not been recorded from any of these sites for more than 40 years, apart from at Dom (see table below). Therefore it may not survive at many of these sites, given the pressures highland habitat has faced in Cameroon (see below). It is not ubiquitous within its range. Despite intensive inventory work, it was not found at Bali Ngemba (Harvey *et al.* 2004).

Unusual in the genus for its high altitude habitat, large leaves, and long brown tomentum.

Habitat: submontane forest, near water?; c. 1000–1630m alt.

Threats: no data are available for the sites near Tibati or Bouar, but at Bamenda, Mbouda and Dom, all within the Bamenda Highlands, there has been extensive loss of forest, especially submontane forest, with 93–96.5% having been lost in total from the original state. Between 1987–1995 alone, 25% of forest cover was lost in the Mt Oku & Ijim Ridge areas of the Bamenda Highlands, based on satellite imagery studies by Moat (Cheek *et al.* 2000: rear cover) (Cheek *et al.* 2000: 49–50.) At Dschang in the Bamenda Highlands, forest loss has been more extreme.

Management suggestions: efforts should be made to rediscover the species in surviving submontane forest patches within its range and then to explore protection of these sites. The highest priority is continued protection of the community forest at Dom, the only known site for the survival of this species where a search should be made for the original plant and where efforts should be made to find more plants.

This species could be propagated and included in forest replanting schemes in the Bamenda Highlands by NGOs such as ANCO, but the preferred proximity of the species for wet ground, if confirmed, should be respected.

Note: Since the original three Ledermann specimens made in 1908–09 at Dschang, Tibati, Mbanti, were destroyed at B, and no illustration is available, our identification of this taxon (by Tchiengue, confirmed by Cheek) has depended on the Mildbraed specimen made in 1914, determined at B, now held at K.

Specimen	Site	Date
Ledermann 1569	Dschang	1908–1909
Ledermann 2405	Tibati	1908–1909
Ledermann 2311	Tibati-Mbanti	1908– 909
Mildbraed 9358	Buar = Bouar, c. 1000m	Fl. May 1914
Meurillon in CNAD 325	Dschang, 1300m	Fr. 14 June 1966
Meurillon in CNAD 940	Mbouda, Plaine de Bagam 1350m	Fl. 5 Sept. 1967
Daramola in FHI 40496	Bamenda town, G. Hospital	Fl. 20 Feb. 1959
Cheek 13469	Dom Forest, Bamenda Highlands, 1630m	Fr. 25 Sept. 2006

Table of *Antidesma pachybotryum* collections

GUTTIFERAE (M.Cheek)

Psorospermum aurantiacum Engl.
VU B2ab(iii)

Range: Nigeria (Obudu Plateau (5 collections)); Cameroon (NW Province: Bambui; Bamenda, Kumbo-Oku; Bafut Ngemba; Bali Ngemba, Dom (15 collections). W Province: Kounden (1 collection)).

This tree or shrub is distinctive for the dense orange-brown hairs on the lower surface of the leaf, the upper surface a contrasting glossy-black when dried. It appears confined to the Bamenda Highlands, with outliers in the adjoining Obudu Plateau and Bamboutos Mts (Kounden).

The assessment above is updated from that in Cheek in Harvey *et al.* (2004).

Habitat: edge of gallery forest; 1500–1800m alt.

Threats: dry season grassland fires, usually set by man, burn into the montane and submontane forest in the Cameroon Highlands, reducing its area. It is possible, even likely, that *P. aurantiacum* by the nature of its habitat, has some resistance to fire and may even benefit from occasional fires. However, the current frequent and intense fires may affect individuals adversely. Conversion of forest to farmland, by contrast, is an undoubted threat. Over 25% of forest in one sample area of the Bamenda Highlands was lost in the 1980s–1990s (Moat in Cheek *et al.* 2000). 50% was lost in another over 15 years (Baena, this volume).

Management suggestions: research to explore the effect of different fire regimes on this species is advised. In the short term, the highest priority is to re-find individuals at the known sites and to seek means to protect these. While almost all natural forest at Bafut Ngemba F.R. has disappeared already, Bali Ngemba F.R. still remains fairly intact and is a good prospect for the conservation of *P. aurantiacum*. However, Dom, with three collections, may represent the greatest concentration of this species in Cameroon.

LEGUMINOSAE – MIMOSOIDEAE (B.J. Pollard, modified by M. Cheek)

Newtonia camerunensis Villiers
CR A2ac+4ac
Range: endemic to the Bamenda Highlands and Bamboutos Mts of Cameroon.

This species was first assessed as CR A1c in Cheek *et al.* (2000), when it was known from only five collections made between 1932-1974 from locations where it is now thought to be absent so at that time was considered to be possibly extinct. The current assessment is maintained from that made by Pollard in Harvey *et al.* (2004).

Much excitement resulted when, in April-May 2002, we rediscovered the species, at Bali Ngemba Forest Reserve and at Laikom below the Ijim Ridge. In May 2002 detailed survey work at Laikom had resulted in two fruiting collections: *Ghogue* 1401 & *Pollard* 1097. Identifications were confirmed by B. Mackinder at Kew. At the Akwamofu sacred forest (Laikom) a large fruiting tree was seen, to 30 m and dbh c. 2.5m, and on the way down the path as far as the bridge across the stream 6 very large trees were seen, with dbh of up to 3m or more. Many seedlings were found, including c. 20 within 5m radius of the tree collected under *Pollard* 1097. The largest tree observed was being smothered and strangled by *Schefflera abyssinica* (Araliaceae), and the second largest tree with two juvenile *Schefflera* growing on the upper part of trunk, resulting in considerable die-back in the crown. The canopy in the Akwamofu sacred forest consists mainly of *Albizia gummifera* and *Newtonia*; the latter having a darker appearance when viewed from the ground, partly due to the darker green colour of the leaflets, but perhaps also the closer arrangement of the leaflets that form a denser covering, allowing less light through to the understorey. Three mature fruiting specimens were seen on the way to the waterfall, and at least four other fruiting trees seen on the way back across to the main path. Seedlings were seen to occur in huge numbers in the understorey, including c. 50 in a $5m^2$ patch, but only perhaps c. 30 plants with dbh between 1 and 5 cm were seen in the whole forest, suggesting poor recruitment to maturity, with even fewer specimens of dbh between 5 and 100cm. Fruit is set abundantly. *Ghogue* 1401 was collected from a further subpopulation recorded in a forest pocket up along the Ijim ridge, which suggests that these fragments are important for the survival of this species. A further subpopulation has been located by Kenneth Tah at Finge (Tah pers. comm. 2002). A sterile sapling collected at Bali Ngemba F.R. during the 25m × 25m plot survey (*BAL*52 in April 2002), was not identified until 2004 at Kew.

Despite exhaustive searching, the flowers of this species remain unknown to science, probably appearing earlier in the season, around March.

Kenneth Tah was first to discover that the forests of Dom have the world's greatest known density of *Newtonia camerunensis,* with possibly a hundred or more stems. It is known from the forest at Kinjinjang Rock and Dom patches 2 (Sagnere)and 3 (Sousi). At all these locations it is frequent

Habitat: submontane to montane forest with *Albizia gummifera, Carapa grandiflora, Syzygium staudtii, Prunus africana, Pterygota mildbraedii;* 1600–2030 m alt.

Threats: the sites at Kilum-Ijim fall outside the boundary of protection, except at Akwamofu where some protection is afforded by the Kom, who do not allow clearance of forest for traditional reasons. Unfortunately this is only enforced on one side of the path, and most of the specimens of *N. camerunensis* appear to occur on the non-protected side. It seems that the proliferation of *Schefflera* poses a threat to mature individuals, from which most of the seed-rain originates. It is likely that these substantial trees are still used as timber and it may well be exploited elsewhere in its known range. At Bali Ngemba F.R. the threat from forest clearance for timber, firewood and small-scale agriculture is increasing every year (Pollard pers. obs.).

Management suggestions: There is scope to investigate the biology of this species more closely, particularly at Laikom and Dom, to investigate: the dynamics of recruitment to maturity; ecological relationships with *Albizia gummifera*; the morphology of the inflorescences and flowers; identification of pollinators and seed-dispersal agents; cultivation regimes, the causes of mature specimen mortality, including the effects of parasitic and strangler plants. Such studies would greatly assist our conservation efforts.

In recent years, funded by Kew, seedlings of this species have been incorporated into the forest restoration plantings led by ANCO at Dom (Mzeka this volume). However, often wild collected seedlings are used since seed germination in nurseries is reported to have a very low success rate (Tah this volume).

MELASTOMATACEAE (M. Cheek)

Dissotis longisetosa Gilg & Ledermann ex Engl.
VU B2ab(iii)
Range: Nigeria (Obudu (2 collections), Mambilla Plateau (Ngurdje F.R. to Gangirwal (4 collections)); Cameroon (NW Province: Bafut Ngemba F.R. (2 collections); Bali Ngemba F.R. (2 collections); Dom, Tabenken. Adamowa Prov.: Ngaoundere (4 collections). W Province: Bamboutos Mts (2 collections)).

This striking, long-setose *Dissotis* is not easily confused with any other *Dissotis*. *Lock* 84/56, from Togo, 740m, referred to this taxon at K, 'appears to belong to another taxon', it has soft hairs and a subsessile inflorescence.

Usually only 1-3 plants occur at a site: they are not spread uniformly through the grassland habitat. Plants are thought to have an underground rootstock. The absence of this species and so many other Cameroon Highland species of *Dissotis* from Mt Cameroon, is remarkable.

The assessment above is updated from Cheek in Harvey *et al.* (2004).

Habitat: heavily grazed and sometimes burnt submontane/montane grassland; 1600–2100 m.

Threats: conversion of grassland to tilled land. The rootstocks may be vulnerable to predation.

Management suggestions: ecologically this appears such an undemanding species that it may be expanding its range since, on the whole, burnt, heavily grazed grassland appears to be increasing within its range! Known from only eight sites, the factors accounting for its rarity are unknown and a detailed study of its ecology is desirable.

It may be that the species is especially vulnerable to cattle trampling. Further surveys may reveal more sites, so downgrading the threat level to this taxon.

MELIACEAE (M. Cheek)

Entandrophragma angolense (Welw.) C.DC.
VU A1cd

Entandophragma angolense is one of the five African internationally-traded timber species of the mahogany family that were listed as Vulnerable by Hawthorne (Hawthorne 1997, www.redlist.org, in IUCN 2003) using the 1994 criteria of IUCN. They all have a wide range in Africa and were they reassessed in this book, without reference to their use as timber, they would probably be downlisted. Hawthorne cites over-exploitation, poor levels of regeneration, fire damage, and slow growth to support his assessments of these species. This text is taken from Cheek in Harvey *et al.* (2004).

MYRICACEAE (M.Cheek)

Morella arborea (Hutch.)Cheek
(syn. *Myrica arborea* Hutch.)
VU A2c+3c

Range: Bioko (Moka area, 4 coll.); Cameroon: Mt Cameroon (two sites, 6 coll.), Mwanenguba (6 coll.), Bamboutos-Djuttitsa, Bamenda Highalnds, Djottin, massif Mbam, Ijim Ridge, Dom (1 coll.).

Morella arborea is one of the few montane tree species restricted to the Cameroon Highlands. Within this area it has wide range, but is generally rather rare. It does not grow in dense forest, but outside it, in rocky grassland areas, appearing as stunted, gnarled, aged trees. The thick bark of the trunk suggests fire resistance. The aromatic, simple, oblong leaves are distinctive. This text is modified from that in Cheek *et al.* 2004: 170.

Habitat: submontane and montane forest edge, or isolated in grassland; 1300-2400m alt.

Threats: clearance of trees for agriculture (e.g. by fire for both pastoral and tilled land) and wood, particularly firewood. The assessment that 30% of the habitat of *Morella arborea* will be lost over the next hundred years, mainly in the Mwanenguba-Bamenda Highlands sector of the Highlands, is maintained here. In addition it is estimated that this percentage loss has already occurred over a similar duration in the same sector.

Management suggestions: research is needed on the ecology and demography of this species.

MYRSINACEAE (M. Cheek)

Ardisia dom Cheek
CR D1

Range: Cameroon (NW Province: Dom, two sites).

This 1.5-2m tall understorey forest shrub is distinct in its 2cm long peduncles, ridged stems, and leaves with conspicuous, large and dense black dots within the blade. More information is available within this volume, where *Ardisia dom* is described as new to science.

Only two individuals, at different sites, are known for this species. One site is unprotected and vulnerable to clearance.

Habitat: submontane forest; 1600–1830m.

Threats: on current data this species is so rare that even low-level human activity in the forest habitat, such as path cutting could easily threaten the survival of this species.

Management suggestions: seed of *Ardisia dom* should be collected from the mother plants, and seedlings raised for incorporation in the existing forest restoration programme at Dom. All forests at Dom should be resurveyed to target *Ardisia dom* and other of the rare species discovered during the completion of the Dom checklist, so that we have a more accurate idea of how many individuals there are, and what the regeneration levels. Such data is needed to inform development of a management plan. A poster campaign featuring this species might heighten awareness of its rarity and increase the chances of its survival.

MYRTACEAE (M. Cheek)

Eugenia gilgii Engl. & Brehm.
CR A2b,c

Range: Nigeria (Mambilla Plateau (13 collections)); Cameroon (Bamboutos Mts (3 collections), Bamenda Highlands (7 pre-1996 collections, 4 recent collections at Ijim, Mbingo, Bali Ngemba and Dom), Ngaoundere (1 collection)).

The assessment above was made in Cheek *et al.* (2000) in IUCN (2003) (www.redlist.org) and maintained in Cheek in Harvey *et al.* (2004). That assessment is again maintained here.

Described in 1917 from *Ledermann* 2131 (NW Cameroon, Tapare to Riban, gallery forest, 1300m, Jan. 1909). Although many herbarium collections of *Eugenia gilgii* exist, it seems highly threatened by the fact that:

a) its natural habitat has been almost completely destroyed
b) what is left is disappearing rapidly and
c) very few individuals are in protected areas.

Nonetheless, this was probably once a relatively common species in the Bamenda Highlands. Apart from two widely separated trees in the Anyajua area and small populations at Laikom, Bali Ngemba and Dom, the only trees of *Eugenia gilgii* that I have seen are in an extensive subpopulation of at least 50 trees discovered in 1999 on the path to Mbingo 'Back Valley'. This may be the main refuge for the species in the Bamenda Highlands, although it has no formal protection. The relatively numerous collections from the Mambilla Plateau were all made over 23 years ago. According to Hazel Chapman (pers. comm.) the situation with regard to habitat destruction in Nigeria over this time has been no better than it has been in the Bamenda Highlands.

Habitat: drier lower montane forest, often at edges; (1200–)1500–2000m alt.

Threats: clearance of forest for wood and land for agriculture.

Management suggestions: the possibility of protecting the subpopulation at Mbingo should be investigated. Seed should be collected from the mother plants, and seedlings raised for incorporation in the existing forest restoration programme at Dom

RUBIACEAE (M. Cheek)

Chassalia laikomensis Cheek
CR A2b,c

Range: Nigeria (Mambilla Plateau (1 collection)); Cameroon (Mwanenguba (1 collection), Bamenda Highlands (several sites: Bali Ngemba, Ijim and Dom)).

The assessment above was made in Cheek & Csiba (2000), listed as having been assessed by Cheek *et al.* (2000) in IUCN (2003) (www.redlist.org). That assessment was maintained in Cheek in Harvey *et al.* 2004 and is maintained here also. A new location, Dom, is added here to its range.

This is a forest understorey shrub, probably long-lived

Habitat: montane evergreen forest; 1650–2000(–2400)m alt.

Threats: about 95% of the original forest cover of the Bamenda Highlands has been lost to e.g. agriculture (Cheek *et al.* 2000; Cheek & Csiba 2000) and there have been similar losses at Mambilla and Mwanenguba. In the Dom area 50% of forest cover was lost in the 15 years 1988-2003 (Baena this volume).

Management suggestions: more information is needed on the numbers of individuals at the known sites and levels of regeneration. Enforcement of existing protected area boundaries would help protect a significant portion of the surviving population.

Oxyanthus okuensis Cheek & Sonké

CR A2b,c

Range: Cameroon, N.W. Province, Mt Oku and the Ijim ridge, and Dom.

Dom has the world's largest known population of this magnificent forest understorey shrub. At least 20 plants were counted at Dom in Sept. 2006, and there are probably many more than that present. Given that 50% of forest in this part of the Bamenda Highlands was lost 1988-2003 (Baena, this volume), it is likely that more than 80% of the habitat of this species has been destroyed over three generations.

This species was first collected, in fruit, by Duncan Thomas from forest near Lake Oku in Feb. 1985 (*Thomas* 4377) and mistakenly identified as *Oxyanthus formosus* (Thomas in McLeod 1986: 62). It was suspected of being new to science once it was recollected in the 1996 survey (Cheek *et al.* 1997). Thought to be restricted to the forest immediately around Lake Oku (where seven plants are known, Kemei *pers. comm.*), until c. four sterile treelets were seen in the reconnaissance to the Ajung Cliff (*Cheek* 10103) led by DeMarco in November 1999. It is notable that surveys in other areas of forest on Mt Oku and the Ijim ridge have not located this species. However, according to Kemei (*pers. comm.*), there is a third site (a single plant) for this species near Ntum at Ijim. A full description and notes on this species can be found in Cheek & Sonké (2000). The assessment above is updated from that in Cheek *et al.* (2000).

Oxyanthus okuensis is probably most closely related to *Oxyanthus montanus* Sonké which is endemic to Bioko, Mt Cameroon and the Bakossi Mts.

Habitat: understorey of submontane and montane evergreen forest; 1590–2200m.

Threats: possibly cut for firewood; forest clearance for agriculture. At Dom it appears secure. Both the Lake Oku and Ajung cliff sites are inside the Kilum-Ijim boundary, and so are protected, so long as the boundaries are respected. However, the forest at Ajung was only recently included inside the boundary and there is still evidence that the more accessible part of this forest was still being cleared, possibly as late as early 1999. The forest at Ntum is not protected.

Management suggestions: more information is needed on the number of individuals of *Oxyanthus okuensis* present at the two known sites, and upon levels of regeneration. The plant at the Ntum site should be vouchered. Fertile material is still needed from the Ajung site in order to confirm the specific identity. This species should be looked for in other areas of forest in the Bamenda Highlands. Confusion with other species of *Oxyanthus* is unlikely since this is the only known member of the genus at high altitude in the Bamenda Highlands.

Pavetta hookeriana Hiern var. *hookeriana*
VU A2bc
Range: Cameroon (Mt Cameroon (numerous collections), Mwanenguba (c. 3 collections), Bamenda Highlands (numerous collections at several sites: Bali Ngemba, Kilum-Ijim and Dom).

It is estimated that over 30% of the habitat of this woody forest understorey species has been lost in the last century.

This is the highest altitude Pavetta known W of the Congo basin. Generally a good indicator of montane forest at 2000m, its occurrence of altitudes of 1500-1600m at Bali Ngemba and Dom is puzzling.

The assessment above is updated from that in Cheek *et al.* (2004) and Cheek in Harvey *et al.* (2004).

Habitat: montane forest; (1500-)1900–2000(–2400)m alt.

Threats: secure from threat at Mt Cameroon, *P. hookeriana* is threatened by forest clearance for agriculture and wood throughout the extensive Bamenda Highlands, probably once the main area for the species. Study of one large area in the highlands between 1987–1995 showed that 25% of the surviving forest was lost (Moat in Cheek *et al.* 2000) and elsewhere 50% was lost 1988–2003 (Baena, this volume).

Management suggestions: improved policing of existing forest reserve boundaries could prevent extinction of this species in the Bamenda Highlands, where its survival is precarious, except at Kilum-Ijum. At Mwanenguba and Bamboutos Mts (presence inferred) it may not survive for much longer. The species is most secure at Mt Cameroon, where the narrowly endemic variety *pubinervia* also occurs.

Psychotria moseskemei Cheek
CR A2bc
Range: Nigeria, Chappal Hendu, Chappal Waddi; Cameroon: N.W.Province: Bamenda Highlands: Ijim at Laikom and Belo (Zitum Rd); Kilum at Elak (Upkim Forest); W Province, 10km W of Bangwa.

It is estimated that over 80% of the habitat of this forest understorey shrub species has been lost in the last century. Its habitat in the Dom area has seen a 50% reduction in the 15 years 1988–2003 (Baena, this volume). All known sites for the species have been under similar pressure and it probably does not survive in all of them. *Psychotria moseskemei* was listed as *Psychotria* sp. nov? in Cheek *et al.* (2000:73), where detailed location data are available of the then known sites. The absence of the species from Bali Ngemba is notable. The assessment above is updated from that in Cheek & Csiba (2000).

The red midrib and secondary nerves, and the glabrous, circular domatia help identification of this species in the field and to separate it from *Chassalia laikomensis* which often grows with it.

Habitat: submontane and montane forest; 1600–2100m alt.

Threats: the Bamenda Highlands is probably the main area for the species. Study of one large area near Mt Oku in the highlands between 1987–1995 showed that 25% of the surviving forest was lost (Moat in Cheek *et al.* 2000) and elsewhere, at Kejojang (Kejodsam), 50% was lost 1988-2003 (Baena, this volume).

Management suggestions: seed of this species should be collected and seedlings incorporated in areas where forest has been restored. Probably this species would not establish where no forest canopy yet exists, nor would it compete with grasses. Existing sites for this species should be secured and protected and populations monitored. As with most tropical shrubs, almost nothing is known about the life-cycle of this species, so this should be studied if resources allow.

SAPINDACEAE (M. Cheek)

Allophylus bullatus Radlk.

VU A2b,c

Range: Príncipe & São Tomé; SE Nigeria; Cameroon; Mt Cameroon, Mt Kupe, Bali Ngemba, Mt Oku and Ijim Ridge, Dom.

This understorey tree of upper submontane to montane forest, while secure on Mt Cameroon and on Mt Kupe, has lost large tracts of its habitat in recent decades in the Bamenda Highlands. Over 30% of its overall habitat is estimated to have been lost in the last 100 years. This taxon has also been assessed for Cheek *et al.* (2004) and Cheek in Harvey *et al.* (2004). Most collections of this species are from Mt Cameroon and from Kilum-Ijim (Mt Oku and the Ijim Ridge) where it appears commonest. *Allophylus bullatus* is much rarer at lower altitudes where the bullate characteristic is often absent. Only 1-2 specimens are known each at Bali Ngemba and Dom. This species is distinctive in having white hairy domatia in the axils of tertiary nerves on the secondary nerves.

Habitat: upper submontane and montane forest; 1600–2400 m alt.

Threats: clearance of forest for agriculture and wood, particularly in the Bamenda Highlands of Cameroon, once probably the main habitat for *A. bullatus*. Study of one area here, the surroundings of Mt Oku (Moat in Cheek *et al.* 2000) showed that 25% of forest was lost between 1987–1995 and elsewhere, at Kejojang (Kejodsam), 50% was lost 1988–2003 (Baena, this volume)..

Management suggestions: improved policing of the existing protected areas could secure the future of this species.

Allophylus ujori Cheek

EN B2ab(iii)

Range: Nigeria (Mambilla-Gembu, Mayo Naga-Njawe (1 collection each)); Cameroon (Kebo (unlocated precisely), Bali Ngemba F.R. (3 collections), Dom (2 collections)).

This 6m forest understorey tree is unusual in its genus for being spiny. This species was confused with *Allophylus conraui* in FWTA and was misnamed as such in our Bali Ngemba checklist (Cheek in Harvey *et al.* 2004). However, that species is a non-spiny 1m shrub of lowland forest.

The assessment above is updated from that in Cheek in Harvey *et al.* (2004).

Habitat: evergreen forest; 1400-1600 m alt.

Threats: forest clearance for agriculture (e.g. Bali Ngemba F.R. where farming has increased inside the forest year by year) and in the Mambilla plateau area of Nigeria (H. Chapman pers. comm.). At Dom, its forest habitat at one site, near Kinjanjang Rock, remains outside the protected areas.

Management suggestions: the status of forest at the two Nigerian sites needs to be ascertained. This would best be done by reference to Chapman and Olsen, who are conducting surveys in the area. In Cameroon, survey work should focus on Bali Ngemba and Dom where several individuals of *A. ujori* have been seen recently (pers. obs.). Seed of this species should be collected and raised for incorporation in forest restoration plantings such as that at Dom.

A definitive enumeration of the subpopulation, and assessment of regeneration and demography, would provide the data needed to formulate a management plan for this species.

Note: this species was mistakenly named as *Allophylus conraui* Gilg ex Radlk. in the Bali Ngemba checklist.

STERCULIACEAE (M. Cheek)

Dombeya ledermannii Engl.

CR A2b,c

Range: Nigeria (Mambilla Plateau (4 collections), Jos Plateau (2 collections)); Cameroon (Bamenda Highlands (7 collections)).

It is easily counted since, in April, trees are clearly visible from a distance on account of their white flowers. Flowering can occur in November (pers. obs.). The habitat of this species is highly threatened in all its known localities. Seyani (1982) reports that *D. ledermannii* is characteristic of forest edges and the early stages of forest succession, but that on exposed rocky slopes is a normal component of stunted, more open forest. In the Bamboutos Mts, it is propagated by cuttings and planted to form hedges (*Letouzey* 201, cited by Seyani 1982). This tree has previously been assessed in Cheek *et. al* (2000 and 2004) and in Cheek in Harvey *et al.* 2004. The only change in the assessment has been the addition of Dom to the range.

Habitat: woodland; (700–)1220–1980m alt.

Threats: clearance for agricultural land, over-exploitation for bast fibre.

Management suggestions: more information is needed on the extent, distribution and threats to *D. ledermanniii* within the Bamenda Highlands which appears to be the centre of distribution of the species.

MONOCOTYLEDONAE

CYPERACEAE (I. Darbyshire, B.J. Pollard & M. Cheek)

Bulbostylis densa (Wall.) Hand.-Mazz. var. *cameroonensis* Hooper

VU B2ab(iii)

Ran ge: Cameroon (SW Province: Mt Cameroon (1 site, 2 collections); Mt Kupe (1 collection). Littoral Province: Mt Mwanenguba (1 collection). NW Province: Bali Ngemba F.R. (1 collection), Dom (1 collection)).

Until the 1990s, this variety was known only from the type collection (*Mann* 1360b) on Mt Cameroon, believed to be from the Mann's Spring area. It was then rediscovered there in 1992 (*Thomas* 9407), and has subsequently been recorded at other montane grassland sites. Although the extent of occurrence is now known to be much greater than when first assessed in Cable & Cheek (1998), the five known sites are highly isolated and limited in size, thus they are still vulnerable to local stochastic change, such as lava flow on Mt Cameroon or natural fires, thus the vulnerable status is still valid. Care must be taken not to confuse this plant with the sympatric typical variety.

The assessment above is updated from that in Cheek *et al.* (2004) and Cheek in Harvey *et al.* (2004).

Habitat: upper submontane and montane grassland; 1800–3000m alt.

Threats: see above.

Management suggestions: more data on the size of each population are required to assess further its vulnerability to local stochastic events. Further botanical inventory work on montane grassland sites in e.g. the Bamenda Highlands, or Mt Nlonako in Littoral Province, may reveal further populations, in which case the conservation status would be downgraded.

Coleochloa domensis Muasya & D.A. Simpson *ined.*
CR D1
Range: Cameroon, NW Province: Dom, two sites.
Tropical Africa's only obligate epiphytic sedge was discovered by Dr Muasya while identifying specimens of this family (collected two years previously) at Kew in late 2008. So far, it is unique to Dom
Habitat: submontane forest; 1600-1830 m alt.
Threats: this species, currently, is known from only three trees
Management suggestions: We assess this plant as Critically Endangered (CR) using the IUCN criteria (IUCN 2001) since it is only known from two sites , one of which is under threat of forest clearance due to agricultural pressure. Overall it is estimated that as much as 96.5% of the original montane (including submontane) forest of the Bamenda Highlands has been lost (Cheek 2004). GIS-based studies within another area of the Bamenda Highlands show that the loss of forest cover continues, with 25-30% lost in the area around Mt Oku between 1987-1995 (Moat 2000). Surveys in other surviving submontane forest fragments of the Bamenda Highlands carried out by British and Cameroonian scientists in recent years have not yet revealed further sites for this species. The Apiculture and Nature Conservation Organisation (ANCO), a non-governmental forest conservation organisation based in Bamenda, has alerted the population of Dom, which manages the forest fragment inhabited by *Coleochloa domensis*, to the existence of this rare species and its conservation importance.

GRAMINEAE (M. Cheek)

Panicum acrotrichum Hook. f.
VU B2ab(iii)
Range: Bioko (El Pico), Nigeria (Obudu Plateau, Mambilla Plateau), Cameroon (SW Province: Mt Cameroon (Idenau waterfall, Mann's Spring. above Buea); NW Province: Mt Oku (1 collection) and Dom; Adamaoua Prov. (fide van der Zon).
This is an unusual species of *Panicum* in that it is restricted to upper submontane-montane forest (mainly 1300-2300m alt.), and does not occur in open grassland. As such its habitat is under threat, e.g. 50% of such forest in the neighbourhood of Dom was cleared in the 15 years 1988-2003 (Baena this volume). Since most grasses have a generation time of only 1 or 2 years, this species cannot here be assessed under criterion A, but under B. Since nine sites are known, with an AOO of 36km² (calculated at 4km²) per site, and past, current and future habitat loss, it rates the above assessment.
This grass, while never described as frequent or common, does not seem to be especially rare within its habitat. 23 specimens are incorporated at K at present.
Habitat: (submontane) and montane forest; (700-) 1550–2350m alt.
Threats: see above.
Management suggestions: continued protection at such sites as Mt Cameroon and Mt Oku will ensure the survival of this species.

ORCHIDACEAE (B.J. Pollard)

Habenaria nigrescens Summerh.

VU A2c; B2ab(iii)

Range: Nigeria: Obudu Plateau (1 coll.); Cameroon: Bafut Ngemba F.R. (4 coll., 1 site), Mt Neshele, 10km ESE Bamenda (1 coll.), Banyo, Mayo Tankou (1 coll.), Mwanenguba (1 coll.), Dom (1 coll.).

The stronghold of this highly localised taxon was previously the Bafut Ngemba Forest Reserve in NW Province, Cameroon, from where the type specimen was collected (*Daramola in FHI* 41568, July 1959), and from where 3 other collections were made between 1959 and 1975. Outside of the Bamenda Highlands, it is known from a single collection in 1973 from the Obudu Plateau of Nigeria (*Hall* 2952), and from a 1971 collection from Mwanenguba, Littoral Province, (*Leeuwenberg* 8473). An unconfirmed record from 1967, made near Ngaoundéré, Adamawa Province, is recorded at a considerably lower altitude (1100m) and in drier climatic conditions to the other known collections, the specimen being labelled as '*Habenaria* sp. aff. *nigrescens*' and likely refers to a separate entity (*Meurillon in CNAD* 844). This assessment is modified from that of Cheek *et al.* 2004.

Habitat: montane grassland; c. 1700–2300m alt.

Threats: a significant reduction in both the quality and area of natural habitat has been recorded in the Bafut Ngemba F.R. (Pollard, *pers. obs.* 2002), and it is likely that this will have had deleterious effects on all plant taxa, irrespective of habitat preferences. Similarly, high pressures from human activities occur in the montane sites at Obudu, Nigeria, and Mwanenguba, Cameroon.

Management suggestions: a survey of this taxon should be made at the Bafut Ngemba Forest Reserve and protective measures should be put in place here if it remains extant.

BIBLIOGRAPHY

Baena, S., Moat, J. & Forbosch, P. (in press). Monitoring Vegetation Cover Changes In Mount Oku And The Ijim Ridge (Cameroon) Using Satellite And Aerial Sensor Detection. In van der Burgt, X., van der Maesen, J. & Onana, J.-M. (eds). Systematics of African Plants. Proceedings Of The 18th AETFAT Congress, Yaoundé, Cameroon. Royal Botanic Gardens, Kew.

BirdLife International (2003). Birdlife's Online World Bird Database: The Site For Bird Conservation. Cambridge. Available at http://www.birdlife.org (accessed July 2009).

Brummitt, R.K. (1992). Vascular Plant Families And Genera. Royal Botanic Gardens, Kew, U.K. 804 pp.

Brummitt, R.K. & Powell, C.E. (eds) (1992). Authors Of Plant Names. Royal Botanic Gardens, Kew, U.K. 732 pp. Available at http://www.ipni.org/ipni/authorsearchpage.do (accessed July 2009)

Cable, S. & Cheek, M (1998). The Plants Of Mount Cameroon, A Conservation Checklist. Royal Botanic Gardens, Kew, UK. lxxix + 198 pp.

Cheek, M. & Csiba, L. (2000). A New Species And A New Combination In *Chassalia* (Rubiaceae) From Western Cameroon. Kew Bull. 55(4): 883—888.

Cheek, M. & Sonké, B. (2000). A New Species Of *Oxyanthus* (Rubiaceae-Gardiniinae) From Western Cameroon. Kew Bull. 55(4): 889—893.

Cheek, M., Onana, J.-M. & Pollard, B.J. (2000). The Plants Of Mount Oku And The Ijim Ridge, Cameroon, A Conservation Checklist. Royal Botanic Gardens, Kew, UK. iv + 211 pp.

Cheek, M., Pollard, B.J., Darbyshire, I., Onana, J.-M. & Wild, C. (eds) (2004). The Plants Of Kupe, Mwanenguba & The Bakossi Mts, Cameroon, A Conservation Checklist. Royal Botanic Gardens, Kew, UK. iv + 508 pp.

Collar, N.J. & Stuart, S.N. (1988). Key Forests For Threatened Birds In Africa. International Council For Bird Preservation, Monograph No. 3. ICBP, Cambridge, UK. 102 pp.

Convention on Biological Diversity (2002). Convention On Biological Diversity: Text And Annexes. United Nations Environment Programme, Montreal, Canada.

Cook, F.M. (1995). Economic Botany Data Collection Standard. Royal Botanic Gardens, Kew, UK. 146 pp.

Courade, G. (1974). Commentaire Des Cartes. Atlas Régional. Ouest I. ORSTOM, Yaoundé.

De Wit, H.C.D. (1958) Revision Of *Afrardisia* Mez (Myrsinaceae). Blumea, suppl. IV: 241-262.

Goetghebeur, P. (1998). Cyperaceae. In: K. Kubitzki (ed.), The Families And Genera Of Vascular Plants 4. Springer, Berlin.Halliday, P. (1984). Myrsinaceae. Flora of Tropical East Africa, Balkema, Rotterdam.

Govaerts, R., Simpson, D.A., Bruhl, J., Egorova, T., Goetghebeur, P. & Wilson, K.L. (2007). World Checklist Of Cyperaceae. Royal Botanic Gardens, Kew.

Haines, R.W. & Lye, K. (1983). Sedges And Rushes Of East Africa. Eastern African Natural History Society, Nairobi.

Harvey, Y. (2007). Towards A Checklist Of The Threatened Forest Remants Of Dom, Bamenda Highlands, Cameroon. The Nigerian Field 72: 101-107 among papers presented at the NFS [Nigerian Field Society] symposium on 'Conservation In South Eastern Nigeria And Cameroon' held at the Royal Botanic Gardens, Kew 2007

Harvey, Y, Pollard, B.J., Darbyshire, I., Onana, J.-M. & Cheek, M. (eds) (2004). The Plants Of Bali Ngemba Forest Reserve, Cameroon: A Conservation Checklist. Royal Botanic Gardens, Kew, UK. iv + 154.

Hawkins, P. & Brunt, M. (1965). The Soils And Ecology Of West Cameroon. 2 Vols. FAO, Rome. 516 pp, numerous plates and maps.

Hepper, F.N. (1963). Myrsinaceae. Flora Of West Tropical Africa, Vol 2. Crown Agents, London

Holmgren, P.K., Holmgren, N.H. & Barnett, L.C. (1990). Index Herbariorum. 8th ed. New York Botanical Garden. 693 pp.

IUCN (1994). IUCN Red List Categories. IUCN, Gland Switzerland. 21 pp.

IUCN (2001). IUCN Red List Categories And Criteria. Version 3.1. IUCN, Gland, Switzerland. 30 pp.

IUCN (2003). The IUCN Red List Of Threatened Species. Available at http://www.redlist.org (accessed July. 2009).

Keay, R.W.J. & Hepper, F.N. (eds) (1954–1972). Flora Of West Tropical Africa, 2nd ed., 3 vols. Crown Agents, London.

Lebrun, J. (1935). Les Essences Forestières Des Régions Montagneuses Du Congo Oriental. Institut national pour l'étude agronomique du Congo Belge, Bruxelles, Belgium. 263pp.

Letouzey, R. (1968). Les Botanistes Au Cameroun. Flore Du Cameroun: 7. Museum National d'Histoire Naturelle, Paris. 110 pp.

Letouzey, R. (1985). Notice De La Carte Phytogéographique Du Cameroun Au 1: 500,000. IRA, Yaoundé, Cameroon.

Mabberley, D.J. (1997). The Plant Book. Second edition. Cambridge University Press, U.K.

Mackay, C.R. (1994). Survey of Important Bird Areas For Bannerman's Turaco Tauraco bannermani And Banded Wattle-Eye Platysteira laticincta In North West Cameroon, 1994. Interim report. BirdLife Secretariat.

MacKinnon, J. & MacKinnon, K. (1986). Review Of The Protected Areas System In The Afrotropical Realm. IUCN Switzerland & UK. xviii + 259 pp.

Mzeka, P.N. (2009). Dom, A Tiny Village In The Bamenda Highlands Of Cameroon Swings Into Prominence. (FAO Africa magazine) Nature & Faune 23(2): 58 (Feb. 2009). See http://www.fao.org/world/regional/raf/workprog/forestry/magazine_en.htm (Accessed July 2009).

Nkembi, L.& Atem, T. (2003). A Report Of The Biological And Socio-Economic Activities Conducted By The Lebialem Highlands Forest Project, South West Province, Cameroon. Environment and Rural Development Foundation (ERuDeF), Cameroon.

Pollard, B.J. (2002). Rediscovery. Kew Scientist 22: 4. Available at http://kew.org/kewscientist/ks_22.pdf (accessed July 2009).

Secretariat of the Convention on Biological Diversity (2002). Convention On Biological Diversity: Text And Annexes. United Nations Environment Programme, Montreal, Canada. 34 pp.

Seyani, J.H. (1982). A Taxonomic Study Of *Dombeya* Cav. (Sterculiaceae) In Africa. D.Phil. thesis, Univ. Oxford, UK.

Stuart, S.N. (ed) (1986). Conservation Of Cameroon Montane Forests. International Council for Bird Preservation, Cambridge, UK.

Taton, A. (1979). Contribution À L'étude Du Genre *Ardisia* Sw. (Myrsinaceae) En Afrique Tropicale. Bull. Jard. Bot. Nat. Belg. 49: 81-120.

Tye, H. (1986). Geology And Landforms In The Highlands Of Western Cameroon: 15–17. *In* Stuart, S.N. (ed) (1986). Conservation Of Cameroon Montane Forests. International Council for Bird Preservation, Cambridge, UK.

WHINCONET   http://www.geocities.com/whinconet (Accessed July 2009).

White, F. (1983). The Vegetation Of Africa: A Descriptive Memoir To Accompany The Unesco/AETFAT/UNSO Vegetation Map Of Africa. Paris, UNESCO. 356 pp. – maps.

World Bank (1993). Ecologically Sensitive Sites In Africa. Vol. 1: Occidental And Central Africa. The World Bank, Washington, USA. 128 pp.

Yana Njabo, K. & Languy, M. (2000). Surveys Of Selected Montane And Submontane Areas Of The Bamenda Highlands In March 2000. Cameroon Ornithological Club, Yaoundé. Cyclostyled.

READ THIS FIRST: EXPLANATORY NOTES TO THE CHECKLIST

Yvette Harvey

Herbarium, Royal Botanic Gardens, Kew, Richmond, Surrey, TW9 3AE, UK

Before using this checklist, the following explanatory notes to the conventions and format used should be read.

The checklist is compiled in an alphabetical arrangement: species within genera, genera within families and families within the groups Dicotyledonae, Monocotyledonae, Pinopsida, Lycopsida (fern allies) and Filicopsida (true ferns), following Kubitzki, in Mabberley (1997: 771–781).

Identifications and descriptions of the species were carried out on a family-by-family basis by both family specialists and general taxonomists; these authors are credited at the head of each account. As a general rule, if two authors are listed, the primary author is responsible for the determinations and the second author for the compiling of the account, including writing of descriptions, distributional data and conservation assessments.

As the incomplete Flore du Cameroun (1963–) is a particularly relevant source of information on the plants of the checklist area, a reference to the volume and year of publication is listed at the head of each family account where available.

The families and genera accepted here follow Brummitt (1992), with recent updates on the Kew Vascular Plant Families and Genera database.

Within the checklist each taxon in the family account is treated in the following manner:

Taxon name

The species name adopted follows the most recent taxonomic work available. Author abbreviations follow the standards of Brummitt & Powell (1992) (www.ipni.org).

Species names not validly published at the time of publication of the checklist are noted as "in press" or "ined." depending upon the extent to which the publication process has advanced, the former indicating that the protologue has been accepted for, but is awaiting, publication and the latter that the species concept is firmly established but that the protologue is not yet submitted for publication.

Not all names listed are straightforward binomials with authorities. A generic name followed by "*sp.*" generally indicates that the material was inadequate to name to species, for example *Crotalaria sp.* (*Leguminosae: Papilionoideae*). Use of "*sp. 1*", "*sp. 2*" and "*sp. A*", "*sp. B*" etc. generally indicate unmatched specimens which may be new to science or may prove to be variants of a currently accepted species; these taxa usually require additional material in order to confirm identity, for example *Rytigynia sp. A of Kupe* (*Rubiaceae*), or new taxa for which sufficient material is available, but are awaiting formal description, for example *Deinbollia sp. 2 of Kupe* (*Sapindaceae*). Unless otherwise stated, or inferred from the distribution, these provisional names are applicable only to the current checklist, thus "*sp. 1 of Dom*" indicates "*sp. 1 of the Dom, Bamenda Highlands, Cameroon checklist*". The use of "*sp. nov.*" is a firm statement that the taxon is new to science but awaiting formal description; sufficient material may or may not be available for this process. A generic name followed by "*cf.*": indicates that the specimens cited should be compared with the associated specific epithet, for example *Harvey* 328 (*Psorospermum* cf. *tenuifolium* Hook.f.) should be compared with *Psorospermum* cf. *tenuifolium* Hook.f. This is an indication of doubt (sometimes due to poor material), suggesting that the taxon is close to (but may differ from) the described taxon. The terms "*aff.*", indicating that the taxon has affinity to the subsequent specific epithet, and "*vel. sp. aff.*", indicating that the specimen refers to the taxon listed or a closely allied entity, are applied in a similar fashion. These uncertainties are generally explained in the taxon's "Notes" section (see below).

Taxon reference

The majority of species referred to within the checklist are found in the 2nd edition of Flora of West Tropical Africa (FWTA: Keay 1954–58; Hepper 1963–72), the standard regional flora. Only species names which do not occur in FWTA are given a reference here; if no reference is cited the taxon name is currently accepted and occurs in FWTA. The references listed are not necessarily the place of first publication of the name; rather, we have tried where possible to use widely accessible publications which provide useful information on that taxon, such as a description and/or distribution and habitat data. The reference is recorded immediately below the taxon name. In the case of scientific journals, we list the journal name, volume and page numbers and date of publication, with recording of volume part number where it aids in access to the publication. In the case of books, we list the surname of the author(s), the book title, the page number for the taxon in question and the year of publication. Journal and book titles are often abbreviated in the interest of economy of space. Several notable publications are:

Fl. Cameroun	Flore du Cameroun (1963–) Muséum National d'Histoire Naturelle, Paris, France & MINREST, Yaoundé, Cameroon
Fl. Gabon	Flore du Gabon (1961–) Muséum National d'Histoire Naturelle, Paris, France.
F.T.E.A.	The Flora of Tropical East Africa (1952–) Crown Agents, London & A.A.Balkema, Lisse, Netherlands.

Synonyms
In some instances, names used in FWTA have been superseded and are thus reduced to synonymy; these are listed below the accepted name, with the prefix "Syn.". Names listed in synonymy in FWTA are not recorded here. Other important synonyms are, however, recorded.

Taxon description

The short descriptions provided for each taxon are based primarily upon the material cited in order to provide the most accurate representation for field botanists working within the checklist area. However, where necessary, they are supplemented by extracted details from the descriptions in FWTA, Fl. Cameroun and the cited taxonomic works. The descriptions are not exhaustive or necessarily diagnostic; rather, they aim to list the key characters to enable field identification of live or dried material, thus microscopic or complex characters are referred to only when they are essential for identification. Where two or more taxa closely resemble one another, a comparative description may be used, by for example stating "Tree... resembling *Strombosia scheffleri*, but ..."; such comparisons are only made to other taxa occurring within the checklist area.

Several abbreviations are used in the descriptions, most notably d.b.h. (referring to "diameter at breast height", being a standard measure of the diameter of a tree trunk), the use of "c." as an abbreviation for "approximately", "±" meaning "more or less".

Habitat
The habitat, recorded at the end of the description, is derived mainly from the field notes of the cited specimens and therefore does not necessarily reflect the entire range of habitats for that taxon; rather, those in which it has been recorded within the checklist area. Habitat information is taken from published sources only where field data are not available, for example, where the only specimens recorded were not available to us, but are cited in FWTA. Altitudinal measurements are derived from barometric altimeters carried by the collectors. These readings have been corrected in the book by Harvey. Altitudinal ranges, listed together with habitat, are generated directly from the database of specimens from the checklist area, and thus do not necessarily reflect the entire altitudinal range known for the taxon. Where no altitudinal data were recorded with the specimens, it is omitted.

Distribution
For the sake of brevity, country ranges are generally recorded for each taxon rather than listing each separate country, for example "Sierra Leone to Uganda" is taken to include all or most of the intervening countries. Only where taxa are recorded from only two or three, rarely more, countries within a wide area of occurrence are the individual countries listed, for example "Sierra Leone, Cameroon & Uganda". For more widespread taxa, a more general distribution such as "Tropical Africa" or "Pantropical" is recorded. Where a species is alien to the checklist area, its place of origin is noted, together with its current distribution. Several country abbreviations are used:

Guinea (Bissau):	former Portuguese Guinea.
Guinea (Conakry):	the Republic of Guinea, or former French Guinea.
CAR:	Central African Republic.
Congo (Brazzaville):	the Republic of Congo or former French Congo.
Congo (Kinshasa):	the Democratic Republic of Congo, or former Zaïre, or former Belgian Congo.

Abbreviations for parts of the country are also used (N: north, S: south (or southern), E: east, W: west, C: central), with the exception of South Africa. Where appropriate, Equatorial Guinea is divided into Bioko, Annobon (both islands) and Rio Muni (mainland), and the Angolan enclave of Cabinda north of the Congo River is recorded separately from Angola itself (south of the river).

In addition to country range, a chorology, largely based upon the phytochoria of White (1983) but with modifications to reflect localised centres of endemism in W Africa, is recorded in square brackets. The main phytochoria used are:

upper Guinea:	broadly the humid zone following the Guinean coast from Senegal to Ghana.
lower Guinea:	separated from Upper Guinea by the "Dahomey Gap", an area of drier savanna-type vegetation, that reaches the Atlantic coast. Lower Guinea represents the humid zone from Nigeria to Gabon, including Rio Muni, Cabinda, and the wetter parts of western Congo (Brazzaville).
Congolian:	the basin of the River Congo and its tributaries, from eastern Congo (Brazzaville) and southern CAR, through Congo (Kinshasa) and to Uganda, Zambia and Angola.
afromontane:	a series of vegetation types restricted to montane regions, principally over 2000m alt.
W(estern) Cameroon Uplands:	a subdivision of the Afromontane phytochorion, used for taxa restricted to the mountain chain running from the Gulf of Guinea islands (Annobon, São Tomé, Principé and Bioko) to western Cameroon and southeast Nigeria.
Cameroon endemic:	for those taxa restricted to Cameroon, a subdivision of Lower Guinea. Taxa endemic to montane western Cameroon are however recorded under W Cameroon Uplands unless they are endemic to the checklist area, when they are listed as a Narrow Endemic (see below).
narrow endemic:	for those taxa restricted to the checklist area, a subdivision of W Cameroon Uplands.

These phytochoria are variously combined where appropriate, for example "Guineo-Congolian (montane)" refers to an afromontane taxa restricted to the mountains of the upper and lower Guinea and Congolian phytochoria. Taxa with ranges largely confined to the Guineo-Congolian phytochorion, but with small outlying populations in wet forest in, for example, west Tanzania or northern Zimbabwe, are here recorded as Guineo-Congolian rather than tropical African, as the latter would provide a more misleading representation of the taxon's true phytogeography. A range of other chorologies are used for more widespread species, including "tropical Africa", "tropical & S Africa", "tropical Africa & Madagascar", "palaeotropics" (taxa from tropical Africa and Asia or some other Old World region), "amphi-Atlantic" (taxa from tropical Africa and S America), "pantropical" and "cosmopolitan". If these terms are used in the distribution, no separate phytochorion is listed.

For some taxa, such as those native to one area of the tropics but widely cultivated elsewhere, the chorology is difficult to define and is thus omitted. Both distribution and chorology are omitted for taxa where there is uncertainty over its identification.

Conservation assessment

The level of threat of future extinction on a global basis is assessed for each taxon that has been fully identified and has a published name, or for which publication is imminent, under the guidelines of the IUCN (2001). Under the heading "IUCN:", each taxon is accredited one of the following Red List categories:

LC: Least Concern   NT: Near-threatened
VU: Vulnerable   EN: Endangered
CR: Critically Endangered

Those taxa listed as VU, EN or CR are treated in full within the chapter on Red Data species, where the criteria for assessment are recorded. Those listed as LC or NT are not treated further in this publication, but it is recommended that further investigation of the threats to those taxa recorded as NT are made. Undescribed taxa, or those with an uncertain determination, are not assessed. In addition, we do not consider it appropriate to assess taxa from the poorly-known genus *Anchomanes* (*Araceae*) for which species delimitation is currently poorly understood and thus for which conservation assessments would be somewhat meaningless; this genus should be revisited once a full taxonomic revision has been completed.

Specimen citations

Specimens from the checklist area are recorded below the distribution and conservation assessment, with the following information:

Location: Dom is a rough indicator as to the actual location, which can be provided by reference to the cited specimen.

Collector and number: within each location, collectors are ordered alphabetically and, together with the unique collection number, are underlined. Only the principal collector is listed here, thus for example, many collections listed under "Onana" were originally recorded as "Onana, Nana F., Harvey, Tah, Corcoran & Johnson Nchianda"; this alteration has been made for the sake of brevity.

Phenology: this information is derived directly from the Cameroon specimens database at Kew and is thus dependent upon recording of such information at the time of collection or at the point of data entry onto the database; if this was not done no phenological information is listed. Collections are recorded as flowering (fl.), fruiting (fr.) or sterile (st.) where applicable.

Date: the month and year of collection is recorded for each specimen; within each collector from each location, collections are ordered chronologically.

The specimens cited are of herbarium material derived from a variety of sources; the chapter on collectors in the checklist area provides further information. In addition, a very few confident sight records of uncollected taxa are also included, also photographic records where no specimen was obtained.

Notes

Items recorded in the notes field at the end of the taxon account include:

- Taxonomic notes, for example in the case of new or uncertain taxa, how they differ from closely related species.
- Notes on the source of specimen data, for example for those specimens recorded in Fl. Cameroun, or for specimens not seen by the author(s) of the family account.

Ethnobotanical information

Local names and uses are listed for each taxon where appropriate; these are derived largely from local residents and field assistants and are reproduced here with their consent. Local names are in the Noni language unless otherwise stated. For each local name or use listed, a source is attributed, usually by reference to the collector and number of the specimen from which the information was derived. The sources of the name or use, where recorded, were the Fon of Dom and Eric Chia. In circumstances where the ethnobotanical information was provided verbally with no attached specimen, or where it is known by the author(s) of the family account, the terms *"fide"*, *"pers. comm."* (personal communication) or *"pers. obs."* (personal observation) are used to attribute the information to a source.

The layout of the information on local uses follows the convention of Cook (1995), listing level 1 state categories of use in capital letters, followed by the level 2 state categories, then the specific use. In order to comply with guidelines set by the Convention on Biological Diversity (2002), the detailed uses of medicinal plants are not listed here; only the level 1 state "MEDICINES" is recorded.

ANGIOSPERMAE
DICOTYLEDONAE

Acanthaceae

I. Darbyshire (K)

Asystasia gangetica (L.) T.Anderson
Straggling herb, to 1m; stems sparsely pubescent; leaves ovate, c. 7 × 3cm, base attenuate then abruptly rounded, apex acuminate, margin subentire, surfaces sparsely pubescent or glabrous; petiole to 1.5cm; inflorescence a 1-sided axillary spike to 14cm, 6–11-flowered; bracts lanceolate, c. 1mm, ciliate; calyx lobes subulate, c. 4mm, pubescent; corolla 2-lipped;, tube 1.3 × 0.8cm, puberulent, white with purple markings at base of throat; stamens 4; ovary and base of style pubescent; fruits 2 × 0.3cm, puberulent. Farmbush & forest edges; 1831m.
Distr.: widespread in palaeotropics, introduced to neotropics [palaeotropics].
IUCN: LC
Dom: Cheek sight record 33 9/2006.

Brillantaisia lamium (Nees) Benth.
Erect perennial herb, to 1.5m; stems tomentose towards apices, hairs to 4mm; leaves ovate, 8.5–11 × 5–6cm, base cordate or truncate, apex long-acuminate, margin subentire, surfaces pilose; petiole to 6.5cm, densely tomentose; inflorescence a lax panicle, glandular-hairy throughout; bracts ovate, 9 × 8mm, apiculate; pedicels to 1.3cm; calyx lobes linear, 0.8cm; corolla bilabiate, 2.5cm long, purple or blue; stamens 2; fruit linear, 3–3.5cm long, glabrous, brown-green. Farmbush & forest margins; 1607m.
Distr.: Guinea (Conakry) to Angola & W Tanzania [Guineo-Congolian].
IUCN: LC
Dom: Cheek sight record 6 9/2006.

Brillantaisia owariensis P.Beauv.
Syn. *Brillantaisia nitens* Lindau
Erect perennial herb or subshrub, to 2.5m; stems robust, glabrous; leaves broadly ovate, 18–21 × 9–11cm, base acute to truncate with a cuneate extension grading into the winged petiole, apex acuminate, margin serrate, surfaces sparsely pilose, cystoliths numerous, visible on upper surface; petiole to 4cm, winged towards apex, tomentose, hairs to 2mm; panicles lax, to 20cm long, many-flowered, glandular-hairy throughout; bracts caducous; bracteoles linear to ovate, c. 7mm long; calyx lobes linear, uneven, longest 1.2cm; corolla bilabiate, c. 3cm, blue-violet with white throat; fruit linear, 2.2cm long, pubescent. Forest & forest-grassland transition; 1735m.
Distr.: Nigeria to S Sudan and W Tanzania [lower Guinea & Congolian].
IUCN: LC
Dom: Harvey 311 2/2006.

Hypoestes forskaolii (Vahl) R.Br.
Fl. Gabon 13: 229 (1966).
Herb, to 1m; stems pubescent; leaves ovate-elliptic, 13–18 × 5–6.5cm, base acute, apex acuminate, margin subcrenulate, sparsely puberulent on lower surface; petiole to 7cm; inflorescence a 1-sided spike to 5cm; bracts paired, partially fused, 0.6 × 0.1cm, apex acute, pubescent, overlapping; corolla bilabiate, c. 1cm long, white with pink spotting; tube twisted through 180°. Farmbush; 1600–1831m.
Distr.: widespread in tropical Africa [tropical Africa].
IUCN: LC

Dom: Cheek 13639 fl., 9/2006; Onana 3619 2/2006.

Justicia striata (Klotzsch) Bullock subsp. *occidentalis* J.K.Morton
Symb. Bot. Ups. 29: 110 (1989); Nord. J. Bot. 10: 390 (1990).
Suberect herb, to 50cm; stems puberulous; leaves elliptic, 4–4.5 × 1.8–2.5cm, base acute, apex barely acuminate, obtuse, margin subcrenulate, finely pilose on both surfaces, cystoliths numerous on upper surface; petiole 1–1.5cm, pubescent; flowers in subsessile axillary clusters of 2–3, puberulent throughout; bracts oblanceolate, 0.8 × 0.2cm; calyx lobes lanceolate, 0.4cm; corolla bilabiate, 0.7–1.2cm long; lower lip trilobed, white with purple markings. Farmbush & villages; 1633m.
Distr.: Ghana, Cameroon & CAR [upper and lower Guinea].
IUCN: LC
Dom: Cheek 13497 fl., 9/2006.

Alangiaceae

J.-M. Onana (YA)

Fl. Cameroun 10 (1970).

Alangium chinense (Lour.) Harms
Tree, 5–18m, glabrescent; leaves alternate, ovate, c. 10 × 6cm, acuminate, obliquely rounded, entire, digitately 5-nerved, scalariform; inflorescences cymose, axillary, c. 20-flowered, 4cm; peduncle 1.5cm; flowers orange, 1.1cm; fruit ellipsoid, 1cm, fleshy. Forest margins and farmbush; 1831m.
Distr.: palaeotropical, excluding upper Guinea [palaeotropics].
IUCN: LC
Dom: Cheek 13641 9/2006.

Annonaceae

G. Gosline (K)

Annona senegalensis Pers. subsp. *senegalensis*
Shrub, or small tree; leaves 5-13 × 3-8.5cm, broadly ovate to ovate-elliptic, subcordate or rounded at base, densely pubescent when young; fruit smooth, yellow, showing outline of carpels by coloured reticulation on the surface. Savanna.
Distr.: Senegal to Cameroon, Congo (Kinshasa), Sudan, East Africa, and to Zimbabwe, South Africa, Madagascar, Comoro & Cape Verde Is. [tropical and subtropical Africa].
IUCN: LC
Dom: Pollard sight record 8 4/2005.

Monanthotaxis littoralis (Bagshawe & Baker f.) Verdc.
Kew Bull. 25: 27 (1971); F.T.E.A. Annonaceae: 99 (1971).
Syn. *Popowia littoralis* Bagshawe & Baker f.
Erect or climbing shrub; leaves oblong-elliptic, rounded to subcordate at base, obtuse at apex, 3.5-10 × 1.5-4.5cm; flowers solitary, extra-axillary, subtended by a conspicuous green heart-shaped bract, 0.5cm; pedicel 2.5-4cm; fruit yellow. Forest; 1450–1640m.
Distr.: Cameroon, Gabon to Uganda [lower Guinea & Congolian].
IUCN: NT
Dom: Nana 140 4/2005; Onana 3612 2/2006.
Local name: Lawe (Nana 140).

Uvariodendron fuscum (Benth.) R.E.Fr.
Tree, to 20m; bark with spicy scent; leafy stems 3mm diam., dark brown, with fairly dense pale brown appressed hairs and

ridges descending half the length of the internode from the petiole bases; leaf blade oblong, to 24 × 5.5cm, acumen 2cm, base acute to obtuse, lateral nerves c. 14 pairs uniting towards the margin, midrib raised above, puberulent, otherwise blade glabrous, quaternary nerves prominent below, areolae with minute raised black spots; petiole 4mm thicker than the stem; flowers and fruits unknown. Forest; 1600m.
Distr.: Mount Cameroon and Bioko [W Cameroon Uplands (montane)].
IUCN: NT
Dom: Cheek 13420 9/2006.
Local name: Lidi Chani (Cheek 13420). **Uses:** ANIMAL FOOD – Fertile Plant Parts – fruits eaten by chimpanzees; FUELS – Fuelwood – firewood (Cheek 13420).
Note: Cheek 13420 is tentatively placed in this taxon by Gosline, xi.2008 (*Uvariodendron sp.* near *fuscum*).

Apocynaceae

M. Cheek (K)

Landolphia buchananii (Hallier f.) Stapf
Agric. Univ. Wag. Papers 92(2): 27 (1992).
Liana, glabrous, lenticellate; leaves pale matt brown, drying green, midrib yellow, papery, elliptic-oblong, c. 8 × 3cm, subacuminate, obtuse, lateral nerves c. 15 pairs, conspicuously tertiary-reticulate; petiole 5mm; panicle terminal, c. 1cm; corolla white, 13mm, twisting to left; fruit globose, 4cm, green with large white circles. Forest; 1831m.
Distr.: SE Nigeria to S Mozambique [tropical Africa].
IUCN: LC
Dom: Cheek 13633 fr., 9/2006.
Local name: Ntooloo (Cheek 13633).
Note: Cheek 13633 equally comes close to one or two other species! (Cheek, xi.2008).

Landolphia dulcis (Sabine) Pichon
Syn. *Landolphia dulcis* (Sabine) Pichon var. *barteri* (Stapf) Pichon
Liana, pilose when young; leaves papery to leathery, ovate to obovate, 4–22 × 1.8–11cm, acumen up to 2.5cm, base cuneate to auriculate, glandular dotted below, lateral nerves 4–7 pairs; petiole 2–17mm; inflorescence axillary, 1–3 per axil, 1–7-flowered, 1.5 × 1.5cm or morel; peduncle 0–5.5mm; calyx lobes to 4 × 2mm; corolla tube 7–20 × 0.9–2mm, tomentose to glabrous; corolla lobes 5–22 × 1.5–4mm; fruit 2–10 × 2–6cm. Forest; 1600m.
Distr.: Senegal to Gabon [upper & lower Guinea].
IUCN: LC
Dom: Cheek 13419 fl., fr., 9/2006.
Local name: Keh-Ngum (Cheek 13419). **Uses:** ENVIRONMENAL USES – Boundaries/Barriers/Supports – used for fences (Cheek 13419).
Note: Cheek 13419 differs in lacking the usual cordate leaf base and shorter blade of this species. Determination may need revision, otherwise not matched in *Landolphia* or related genera (Cheek, xi.2008).

Landolphia landolphioides (Hallier f.) A.Chev.
Liana, 5-30m high; trunk up to 30cm diam., bark fissured; leaves 6-26 × 2.8-10.5cm, coriaceous when dry, paler beneath with pale venation, lateral nerves in 7-16 pairs, obtuse or acuminate with acumen to 1cm long, rounded or acute or slightly angustate at base; petiole 4-17mm long; inflorescence an axillary or terminal cyme, 2.5-6.5 × 1.9-4cm; peduncle 3-15mm long; pedicels 2-8mm long; corolla white, cream or yellow often tinged orange, red, purple or green; tube 8-14mm long; lobes 6.5-17.5 × 2.5-4mm, rounded at apex; fruit yellow, pyriform to globose or ovoid,

3.5-8 × 2.5-7 × 2.5-7cm; seeds irregularly ovoid, 11-21 × 6-13 × 6-8mm. Forest; 1640m.
Distr.: Nigeria to Uganda [lower Guinea & Congolian].
IUCN: LC
Dom: Pollard 1390 4/2005.
Note: Pollard 1390 has leaves that are an excellent match for this species. However, the fruits are supposed to be 5cm long, not 15cm long, so may belong to another taxon, (Cheek, xi.2008).

Oncinotis pontyi Dubard
F.T.E.A. Apocynaceae: 107 (2002).
Climber, with white exudate; stems purple brown, 4mm diam.; leaves opposite, simple, obovate, c. 10.5 × 5.5cm, acumen 0.5cm, base acute, midrib impressed above, lateral nerves 4 pairs meeting near margin, domatia crater like with rim long hairy; petiole 0.7-1.5cm; fruit follicles 3, patent, held in a straight line, each 15 × 1cm, dark brown, densely lenticelled, lenticels pale brown, raised; seed plumes white, 5cm, seedbody 1.5 × 0.3cm, dark brown. Forest, forest edge; 1600m.
Distr.: Cameroon, CAR, Congo (Kinshasa), Ivory Coast, Gabon, Equatorial Guinea, Nigeria, Sudan [Guineo-Congolian].
IUCN: LC
Dom: Onana 3620 2/2006.
Note: the warty fruits of this specimen (Onana 3620), and the acute, not rounded acumen, do not fit this species well so determination may need revising (Cheek, xi.2008).

Rauvolfia vomitoria Afzel.
Shrub or tree, 2–8m; stems greyish-white; leaves in whorls of 3, membranous-papery, elliptic c. 20 × 7cm, subacuminate, cuneate, lateral nerves 12 pairs; petiole 2cm; panicles puberulent, 10cm; corolla white, 8mm; fruit with 2 ovoid berries 8mm. Forest & savanna; 1735–1831m.
Distr.: Senegal to Uganda [Guineo-Congolian].
IUCN: LC
Dom: Dom photographic record 30 2/2006; Cheek sight record 41 9/2006.

Tabernaemontana sp. of Bali Ngemba
Harvey, Y. *et al.*, The Plants of Bali Ngemba Forest Reserve: 84 (2004).
Tree, to 10m, glabrous; leaves to 23 × 9.5cm, elliptic to oblong, acute to subacuminate, cuneate, c. 15-16 pairs of lateral nerves, margin revolute; petiole to 20mm; flowers white; fruit slightly bilobed, deep longitudinal ridges, dark green with yellow spots. Disturbed forest; 1570–1831m.
Distr.: Bali Ngemba F.R., Dom & Fosimondi [W Cameroon Uplands].
Dom: Cheek 13637 fr., 9/2006; Onana 3629 2/2006.
Note: Onana 3629 matches Tadjouteu 414 of Bali Ngemba (previously cited as *Tabernaemontana pachysiphon*), also in bud (Cheek, xi.2008).

Araliaceae

M. Cheek (K) & B.J. Pollard

Fl. Cameroun 10 (1970).

Polyscias fulva (Hiern) Harms
Tree, to 30m; leaves to 80cm long, imparipinnate; leaflets 3-12 paired, coriaceous, to 17 × 7.5cm, lanceolate-ovate; inflorescence of compound racemosely arranged racemes; pedicels to 5mm; petals greenish to creamy white; fruit broadly ovoid to subglobose, 3-6 × 2-5mm, ribbed. Submontane forest on slopes of hills; 1590–1831m.
Distr.: tropical Africa [tropical Africa].

IUCN: LC
Dom: <u>Dom photographic record 29</u> 2/2006; <u>Cheek sight record 36</u> 9/2006.

Schefflera abyssinica (Hochst. ex. A.Rich.) Harms

Deciduous epiphyte or tree, to 30m; bark fissured; petioles to 40cm; leaflets ovate, 15 × 9cm, gradually acuminate, shallowly cordate, serrate, lateral nerves 10 pairs; petiolules 7cm; inflorescence an umbel or short raceme, branches to 40cm; fruits subspherical, to 5 × 5mm. Forest.
Distr.: Cameroon, Congo (Kinshasa), Sudan, Ethiopia, Uganda, Kenya, Tanzania, Malawi, Zambia [afromontane].
IUCN: LC
Dom: <u>Pollard sight record 11</u> 4/2005.

Aristolochiaceae

M. Cheek (K)

Pararistolochia ceropegioides (S.Moore) Hutch. & Dalziel

Adansonia (Sér. 2) 17: 482 (1978).
Climber, 5–15m; stem 8-shaped in section, 1–2cm diam., flexible; leaves leathery, ovate, c. 13 × 8cm, 3-nerved, acuminate, rounded-truncate; inflorescence cauliflorous, 1–2m from ground, 3–8 flowered; flowers 5–7cm; perianth tube pouched at base, pale brown outside; tepals 3, equal, 1–2cm, slightly splayed, throat orange; fruit fleshy, ellipsoid, c. 5–6 × 3–4cm, 5-ridged, base and apex truncate; pedicel 3–4cm. Forest; 1600m.
Distr.: Cameroon & Gabon [lower Guinea].
IUCN: VU
Dom: <u>Cheek 13441</u> 9/2006.
Note: *Pararistolochia preussii* of Kumba, of which the type was destroyed, appears to be continuous with this taxon insofar as perianth length, a key character in Poncy (1978).

Asclepiadaceae

D.J. Goyder (K)

Batesanthus purpureus N.E.Br.

Woody twiner; latex white; leaves opposite, to c. 16 × 10cm, oblong, glabrous, with interpetiolar stipular fringe; flowers in lax axillary panicles; corolla purple, rotate; lobes to c. 18 × 11mm, rounded; corona indistinct, annular; follicles stout, erect, parallel. Forest; 1700m.
Distr.: Cameroon [W Cameroon Uplands].
IUCN: NT
Dom: <u>Corcoran 13</u> 4/2005.

Epistemma cf. *decurrens* H. Huber

Bull. Mus. Natl. Hist. Nat., B, Adansonia, Sér. 4 11(4): 447–452 (1989, publ. 1990). (1990).
Epiphytic shrub; old stems 5mm diam. (dried), purple, glabrous, with copious white exudate, leafy stems brown, matt, white-puberulent; leaves oblong-elliptic, c. 6.5 × 2.5cm, acumen 3mm, lateral nerves c. 6 on each side of the midrib; petiole 6–8mm; inflorescence axillary, nodding, 5-flowered; peduncle 5mm; pedicel 8mm; calyx 5-lobed, 2–3mm; corolla yellow-green, 1 × 1cm, divided by 1/2 to 2/3; corona lobes filiform, densely-hairy. Upper forest edge; 1831m.
Distr.: Cameroon (Bamenda Highlands) [W Cameroon Uplands].
Dom: <u>Cheek 13648</u> fr., 9/2006.
Note: although the Dom specimen lacks flowers, it is very likely to represent only the third known site for *E. decurrens* (CR). The type location is at Banyo and what is probably the

same species was found at Bali Ngemba in 2004 (Darbyshire 378; Harvey et al 2004: 85). The description above is taken from the Bali Ngemba material.

Mondia whitei (Hook.f.) Skeels

Liana, to 4m; white exudate; stems pubescent; leaves opposite, ovate, 14-17.5 × 10-12cm, acumen 1.5cm, base cordate, 5 nerves on each side of the midrib, entire, abaxial surface densely pubescent, adaxial surface sparsely pubescent; petiole (3-)6-8cm; inflorescence a panicle to 15cm; corolla 0.7cm, outside pale green, inside yellow - dark red; 1600m.
Distr.: Guinea (Bissau), Guinea (Conakry), Sierra Leone, Liberia, Ghana, Togo, Benin, Nigeria, Cameroon, Equatorial Guinea, Congo (Kinshasa), Burundi, Mozambique, Malawi, Zambia, Zimbabwe, Angola, South Africa [Guineo-Congolian].
IUCN: LC
Dom: <u>Cheek 13429</u> 9/2006.
Note: Cheek 13429 (known locally as Gow-gow) is tentatively placed in this taxon. The stipular fringe is okay, but the leaf base is atypical (Goyder, iii.2008).

Balsaminaceae

J.-M. Onana (YA) & M. Cheek (K)

Fl. Cameroun 22 (1981).

Impatiens hochstetteri Warb. subsp. *jacquesii* (Keay) Grey-Wilson

Fl. Cameroun 22: 15 (1981).
Syn. *Impatiens jacquesii* Keay
Erect succulent herb, 45cm; leaves alternate; blade narrowly elliptic, to 12 × 3cm, crenate-dentate, with mucros; petiole to 5cm, with decurrent blade; inflorescence terminal, 4-8-flowered; axis highly contracted; pedicels 2.5-3.5cm; flowers pale pink, c. 1cm; petals flat; spur slender, c. 1.5cm. Submontane forest remnants adjacent to farms; 1607m.
Distr.: Guinea (Conakry) to Cameroon [upper & lower Guinea].
IUCN: NT
Dom: <u>Cheek 13400</u> fl., 9/2006.
Local name: Fintambursi (Cheek 13400).

Impatiens mackeyana Hook.f. subsp. *zenkeri* (Warb.) Grey-Wilson

Grey-Wilson C., Impaties of Africa: 221 (1980).
Syn. *Impatiens zenkeri* Warb.
Erect terrestrial perennial, to 1m; stem simple or moderately branched, becoming glabrescent with age; leaves broadly ovate to broadly elliptic-oblong, 5.0-18.5 × 2.5-8.5cm, margin serrate; petiole without stipitate glands; flowers solitary, pedunculate; mauvish-pink or purplish; lateral united petals considerably longer than the lower sepal; ovary glabrous; fruit 17-20 × 5-6.5mm, fusiform, glabrous. Along river margins and moist sites; 1600m.
Distr.: Cameroon, Congo (Kinshasa), Gabon [lower Guinea].
IUCN: LC
Dom: <u>Cheek 13460</u> fl., 9/2006.
Note: Cheek 13460 is a tentative determination (Onana, vi.2009).

Impatiens sakerana Hook.f.

Terrestrial herb, 0.3–1m; leaves in whorls of 3–4 or opposite, ovate-elliptic, c. 7.5–12 × 3.5–4cm, apex acuminate, margin crenate, pale brown-hairy below; racemes short, 3–4-flowered; peduncle c. 4cm; flowers 1.5–3cm long, lip red, gradually narrowing into short spur, greenish-red, swollen at tip. Forest patches.

Distr.: Bioko & Cameroon [W Cameroon Uplands].
IUCN: VU
Dom: <u>Pollard sight record 9</u> 4/2005.

Bignoniaceae

J.-M. Onana (YA)

Fl. Cameroun 27 (1984).

Kigelia africana (Lam.) Benth.

Tree, to 15m, glabrous; leaves opposite, imparipinnate, to 50cm, 5–6-jugate, leaflets oblong-elliptic, 11–20 × 3.5–8cm, apex acuminate, base acute to rounded; petiole 8–25cm; petiolules 3–6mm; panicles pendent, lax, to 40cm, lateral branches to 7cm; peduncle to 20cm; pedicels 1.2cm; calyx cupular, 2.8 × 2.5cm, lobes triangular, 0.6–1cm; corolla campanulate, 6(–8)cm, dark-red; fruit sausage-shaped, 20–50(–100)cm. Forest & farmbush; 1590m.
Distr.: Senegal, Gambia, Mali, Guinea (Conakry), Sierra Leone, Liberia, Ivory Coast, Ghana, Togo Republic, N Nigeria, S Nigeria, Cameroon [tropical Africa].
IUCN: LC
Dom: <u>Onana 3607</u> 2/2006.

Boraginaceae

M. Cheek (K)

Cynoglossum coeruleum A.DC. subsp. *johnstonii* (Baker) Verdc. var. *mannii* (Baker & C.H.Wright) Verdc.

Syn. *Cynoglossum lanceolatum* Forssk. subsp. *geometricum* (Baker & C.H.Wright) Brand
Herb, 0.3-1.2m tall; cauline leaves elliptic to lanceolate, 2-22 × 0.8-5.5cm, sessile or short-petiolate; radical leaves long-petiolate; nutlets with glochidia only on meridian and top surface. Submontane forest, swamp forest, margins of cultivation; 1931m.
Distr.: Cameroon, Congo (Kinshasa), Rwanda, Burundi, E Africa, SE Africa to Zimbabwe [tropical & subtropical Africa].
IUCN: LC
Dom: <u>Cheek 13590</u> fl., 9/2006.

Buddlejaceae

M. Cheek (K)

Nuxia congesta R.Br. ex Fresen.

Syn. *Lachnopylis mannii* (Gilg) Hutch. & M.B.Moss
Tree, 2–8m; bole white, fibrous; stems orange-brown; leaves in whorls of 3, dimorphic: type 1 ovate-elliptic, c. 4 × 2.5cm, apex rounded, base cuneate, serrate, petiole 0.2cm; type 2 c. 11 × 5.5cm, entire, petiole 2cm; inflorescence terminal, 7 × 10cm, dense, many-flowered; flowers white, c. 7mm. Forest & forest edge; 1640m.
Distr.: Guinea (Conakry) to South Africa [Afromontane].
IUCN: LC
Dom: <u>Onana 3634</u> 2/2006.

Campanulaceae

M. Thulin (UPS) & Y.B. Harvey (K)

Lobelia columnaris Hook.f.

Herb, 2m tall, unbranched; stem 2cm diam. at base, glabrous; leaves oblanceolate-oblong, c. 30 × 8cm at base of stem, gradually diminishing in size towards the apical inflorescence, apex acute, margin inconspicuously serrate, softly hairy; inflorescence a densely-flowered, unbranched spike, occupying the apical half of the plant; corolla pale blue, lilac to white, 2-3cm long. Forest edges and grassland; 1650m.
Distr.: Bioko, Cameroon and Nigeria [W Cameroon Uplands (montane)].
IUCN: NT
Dom: <u>Dom photographic record 14</u> 2/2006.

Lobelia rubescens De Wild.

Syn. *Lobelia kamerunensis* Engl. ex Hutch. & Dalziel
Straggling herb, to 45cm; stems triangular with narrowly-winged edges, often rooting at nodes; leaves lanceolate, gradually narrowing upwards on stem, 13–30 × 3–10mm, margin serrate, subglabrous; petioles c. 5mm long; flowers in leafy racemes; pedicels 7–14mm long; bracts to 14mm long; hypanthium 5–6mm long (extending to 7mm in fruit); calyx lobes narrowly triangular; corolla to 10mm long, blue with yellow throat; seeds c. 0.5mm long, brown. Farmbush, grassland; 1780–1815m.
Distr.: Nigeria to Tanzania [lower Guinea & Congolian].
IUCN: LC
Dom: <u>Cheek 13545</u> 9/2006; <u>13567</u> fl., 9/2006; <u>Harvey 308</u> 2/2006.
Note: Harvey 308 (see photograph) is from a damaged plant and is not typical (Thulin, xii.2008).

Capparaceae

O. Sene (YA)

Fl. Cameroun 29 (1986).

Cleome rutidosperma DC.

Fl. Cameroun 29: 57 (1986).
Syn. *Cleome ciliata* Schum. & Thonn.
Herb; branches prostrate, radiating from taproot, c. 30cm; leaves 2.5cm, leaflets 3, 1.5cm; flowers single, erect, 0.7cm; petals 4, 0.5mm, erect; fruit cylindrical, 4cm, glabrous. Roadside; 1600m.
Distr.: pantropical, originating in West Africa [pantropical].
IUCN: LC
Dom: <u>Cheek 13409</u> fl., 9/2006.
Note: Cheek 13409 was identified in the field and has yet to be verified (Harvey, iii.2009).

Ritchiea albersii Gilg & Benedict

Shrub or tree, to 11(-20)m; leaves simple to 5-foliolate; leaflets elliptic or oblong-elliptic, to 20 × 8cm, base cuneate; petiolule 2-5mm long; inflorescence terminal; pedicels 1.5-3.5cm; sepals lanceolate or ovate, 12-20 × 5-10mm; petals 4, to 3.5cm, clawed; gynophore 2-3cm long; fruits cylindrical, to 5 × 2.5cm with 6 longitudinal grooves. Forest; 1600m.
Distr.: SE Nigeria to E Africa [afromontane].
IUCN: LC
Dom: <u>Cheek 13445</u> 9/2006.

Caryophyllaceae

M. Cheek (K)

Drymaria cordata (L.) Willd.

Straggling, slightly viscid herb; stems to 90cm; leaves broadly ovate, 1–2.5cm long; inflorescence a terminal cyme;

flowers few, white. Montane or submontane grassland; 1636m.
Distr.: pantropical [pantropical].
IUCN: LC
Dom: Cheek sight record 23 9/2006.

Celastraceae

M. Cheek (K)

Fl. Cameroun 19 (1975) & 32 (1990).

Hippocratea sp. 1 of Dom
Monopodial tree or young climber, 4m; exudate and scent absent, stems glossy purple-brown, glabrous; leaves opposite, articulate, with a white stipule scar or ridge 3mm long between the petiole bases; blades elliptic-lanceolate, leathery, to 23 × 10cm, acumen 1-2cm, base acute-obtuse, lateral nerves c. 8 pairs, pale brown, quaternary nerves conspicuous, areolae isodiametric, margin subserrate, glands not seen; petioles 1-1.2cm, brown, articulated at base; flowers and fruits unknown. Submontane forest; 1831m.
Dom: Cheek 13640 9/2006.
Note: this specimen (Cheek 13640) probably represents a juvenile canopy liana of the *Hippocratea* alliance but flowers and fruits are needed to identify fully to species and genus (Cheek, vi.2009).

Maytenus buchananii (Loes.) R.Wilczek
Fl. Cameroun 19: 28 (1975).
Small tree; older branches with spines 2–3cm long; leaves elliptic, 3.5–6 × 1.3–2.5cm, toothed, base strongly attenuate; inflorescence an axillary cyme 1–1.5cm long; sepals as long as petals; flower to 7.5mm long; fruit a tri-valved capsule 7–8mm long each valve with 1–2 brown seeds. Forest edge, riverbanks; 1636–1810m.
Distr.: Ivory Coast to Mozambique [tropical Africa].
IUCN: LC
Dom: Cheek 13526 fr., 9/2006; 13656 fl., fr., 9/2006; Corcoran 6 4/2005; Harvey 312 2/2006.

Salacia erecta (G.Don) Walp. var. *erecta*
Climber, to 15m, no resinous threads; branchlets dark brown smooth; leaf blade elliptic 4–15 × 1.5–5cm, 7–10 secondary nerves, margin toothed; petioles to 5mm, channel margins undulate; flowers in sessile axillary fascicles, sometimes on thin shoots appearing like peduncles; pedicels 3–7mm; buds ovoid; flowers c. 5mm diam., yellow; fruits globular, 1–3cm diam., orange-red. Forest; 1550m.
Distr.: Guinea (Conakry) to Zambia [tropical Africa].
IUCN: LC
Dom: Pollard 1382 4/2005.

Combretaceae

M. Cheek (K)

Fl. Cameroun 25 (1983).

Combretum fragrans F.Hoffm.
Fl. Zamb. 4: 120 (1978).
Tree, to 10m; crown rounded; bark creamy(-grey)-brown; leaves 20 × 9cm, tomentose to ± glabrous, lepidote; petiole to 1.5cm; inflorescences axillary spikes to 7cm; rachis pubescent; bracts 1.5mm; flowers sessile, greenish or dirty white; sepals 1 × 1.5mm; petals 2-3 × 1-1.5mm; fruit brown or reddish-yellow, 2.5-3.5 × 2.5-3cm; wings up to 12mm wide. Forest; 1640m.
Distr.: widespread from W Africa to E Africa and S to Zimbabwe [tropical Africa].

IUCN: LC
Dom: Dom photographic record 3 2/2006; Nana 144 4/2005.

Combretum molle R.Br. ex G.Don
Small tree, 5-7(-17)m; bark dark grey to black, rough, reticulately fissured; branchlets with bark peeling in grey fibrous strips; leaves opposite; lamina narrowly elliptic to broadly ovate or obovate, up to 21 × 12.5cm, acute, cuneate, pubescent upper, grey tomentose lower; petiole 2-3mm; inflorescences of axillary spikes, 7-11cm long; peduncles 1-2cm; flowers yellow or yellowish green, fragrant; fruits 4-winged, subcircular to elliptic, 1.3-2.5 × 1.5-2.5cm. Wooded and grassland areas; 1633m.
Distr.: tropical & subtropical Africa, and Yemen [tropical & subtropical Africa].
IUCN: LC
Dom: Cheek 13490 9/2006.
Local name: Fui (Cheek 13490).

Terminalia mollis M.A.Lawson
Tree, 5-13(-20)m; crown rounded; bark dark-grey; branchlets tomentose, grey-brown; leaves 16-37 × 7-19cm, elliptic to obovate-oblong, tomentose; petiole 3.5-5cm; inflorescences axillary spikes, 8-17cm; peduncle 1-2cm; flowers cream-greenish white or pinkish, aromatic; fruit 6.5-12 × 2.5-5.5cm, velutinous. 1633–1640m.
Distr.: W tropical Africa, Congo (Kinshasa), E Africa, Angola & Zambia [tropical Africa].
IUCN: LC
Dom: Cheek 13491 9/2006; Nana 150 4/2005.

Compositae

M. Cheek (K), H. Beentje (K) & D.J.N. Hind (K)

Ageratum conyzoides L.
Annual herb, 0.3–1m tall; cauline leaves ovate, 5 × 2.5cm, apex acute, base acute to truncate, margins serrate; petiole c. 1cm, white pilose; capitula discoid, blue, c. 6mm diam.; numerous in dense terminal aggregations; involucre with phyllaries narrowly oblong, c. 4mm long, green, with two white lines. Farmbush; 1600m.
Distr.: Senegal to Tanzania [pantropical].
IUCN: LC
Dom: Dom photographic record 4 2/2006.

Aspilia africana (Pers.) C.D.Adams subsp. *africana*
Syn. *Aspilia africana* (Pers.) C.D.Adams var. *minor* C.D.Adams
Scrambling shrub or herb, to 2m tall; cauline leaves lanceolate, c. 7.5 × 2.8cm, apex acute, base rounded, margin serrate, upper and lower surfaces scabrid; capitula terminal, single or few, 2.5cm diam., radiate; orange, ray florets c. 6 × 3mm, apex bilobed. Farmbush; 1633m.
Distr.: Senegal to Cameroon [upper & lower Guinea (montane)].
IUCN: LC
Dom: Cheek sight record 15 9/2006.

Berkheya spekeana Oliv.
Shrubby, thistle-like herb to 2.5m; stems few- to many-branched; leaves sessile; blades linear lanceolate to narrowly elliptic, deeply lobed or toothed; capitula solitary or corymbose, with yellow ray florets and orange-yellow disc florets; phyllaries 3-4-seriate; achenes turbinate in alveolate receptable; pappus of numerous scale-like, scarious, setae. Grassland; 1633m.

Distr.: Nigeria, Cameroon, Congo (Kinshasa), Sudan, Ethiopia, Uganda, Kenya & Tanzania [tropical Africa].
IUCN: LC
Dom: Cheek 13492 fl., 9/2006.

Bidens barteri (Oliv. & Hiern) T.G.J.Rayner
Kew Bull. 48: 483 (1993).
Syn. *Coreopsis barteri* Oliv. & Hiern
Annual herb, 0.3-1m tall; stems purple; leaves lanceolate, c. 5 × 1.5cm, apex acute, base rounded, margin serrate, glabrous; petiole 1-2mm long; capitula few, 3cm diam., radiate; rays yellow, clawed, 14 × 4mm; disc florets black; involucre of two whorls of c. 10 phyllaries each, inner broader, black, outer green. Farmbush; 1797m.
Distr.: Ghana to Cameroon [lower Guinea].
IUCN: LC
Dom: Cheek 13553 fl., 9/2006.
Note: Cheek 13553 is much smaller than most other specimens due to shallow soils and poor nutrients perhaps. The lobed leaves are atypical but not unknown for the species. More material needed (Cheek, xi.2008).

Conyza clarenceana (Hook.f.) Oliv. & Hiern
Syn. *Conyza theodori* R.E.Fr.
Perennial stoloniferous herb, to 1.5 m; stems reddish, glabrous or puberulous; leaves pseudopetiolate, oblanceolate to linear; inflorescences a terminal panicle; capitula pedicellate; phyllaries 2-3-serriate; florets many; corollas yellow or pale yellow; achenes flattened, sparsely pilose or glabrous; pappus setae white. Swamp edge; 1780m.
Distr.: Bioko, Mount Cameroon, Congo (Kinshasa), Zambia, Ethiopia, Uganda and Kenya [afromontane].
IUCN: LC
Dom: Cheek 13549 fl., 9/2006.

Conyza pyrrhopappa Sch.Bip. ex A.Rich.
F.T.A. 3: 318 (1877).
Shrub or woody herb, to 3m, pubescent; leaves narrowly elliptic to oblanceolate, acute; capitula campanulate, 0.75-2cm wide; phyllaries linear lanceolate, acute; corollas tiny, white or yellow; achenes hairy; pappus reddish or tawny. Fallow, grassland; 1650–1931m.
Distr.: Nigeria to Sudan, south to Angola & Zambia [lower Guinea & Congolian].
IUCN: LC
Dom: Cheek 13571 fl., 9/2006; Dom photographic record 5 2/2006.

Conyza subscaposa O.Hoffm.
Perennial herb, with a weak flowering shoot from 5-30cm high; leaves forming a basal rosette; corollas dull yellow; phyllaries often purplish. Grassland; 1797m.
Distr.: W Cameroon to Congo (Kinshasa) & E Africa [afromontane].
IUCN: LC
Dom: Cheek 13557 fl., 9/2006.

Crassocephalum bauchiense (Hutch.) Milne-Redh.
Herb, 0.3–1m tall; stems with white crisped hairs; cauline leaves pinnately lobed, c. 7 × 3.5cm, lobes 5, divided almost to the base, margins denticulate; capitula 1cm diam., discoid, corollas blue or purple; involucre of c. 20 inner bracts, as long as capitulum, and c. 20 short, patent linear basal outer bracts 2–3mm long. Forest edge; 1980m.
Distr.: N Nigeria & Cameroon [lower Guinea (montane)].
IUCN: VU
Dom: Pollard 1399 4/2005.
Uses: MEDICINE (Pollard 1399).

Note: Pollard 1399 is the only record of this taxon at Dom. The identification was made in the field and has yet to be verified (Harvey, iii.2009).

Crassocephalum bougheyanum C.D.Adams
Herb, 1-2m tall; cauline leaves simple, ovate, c. 6 × 4cm (lowest leaves sometimes trilobed), apex acute, base truncate to obtuse, margin bidentate; petiole 1-2cm long; capitula discoid, 1-3cm diam., corollas orange, yellow or red. Farmbush or forest edge; 1570m.
Distr.: Mt Cameroon, Bamenda Highlands and Bioko [W Cameroon Uplands (montane)].
IUCN: NT
Dom: Dom photographic record 36 2/2006.
Note: tentative determination, specimen required (Beentje, vi.2009).

Crassocephalum crepidioides (Benth.) S.Moore
Annual herb, to 1m, puberulous or glabrous, branched; leaves deeply lyrate-pinnatifid, sometimes not lobed, 2-25cm long, mostly petiolate; capitula oblong, c. 1cm long, on slender pedicels of 0.5-5cm, in a dense or lax corymbose cyme; phyllaries linear-subulate; corollas yellow or with purplish tips, appearing brownish-red. Weed of farms and waste places; 1659m.
Distr.: Guinea (Conakry) to Mascarenes, naturalised in tropical Asia & Pacific islands [palaeotropics].
IUCN: LC
Dom: Cheek 13535 fl., 9/2006.

Crassocephalum rubens (Juss. ex Jacq.) S.Moore
Syn. *Crassocephalum sarcobasis* (DC.) S.Moore
Annual herb, to c. 1.5m, puberulous; leaves irregularly dentate or lobed, 2-9cm, upper leaves sessile, lower leaves petiolate; capitula terminal, solitary, 1-2cm long, oblong-hemispherical; peduncles 15-30cm; phyllaries linear subulate, glabrous; corollas tiny, purple; anthers included. Fallow & rocky slopes in grassland; 1797m.
Distr.: Cameroon, Sudan to South Africa, Madagascar, Yemen [afromontane].
IUCN: LC
Dom: Cheek 13558 fl., 9/2006.

Crassocephalum vitellinum (Benth.) S.Moore
Herb, 0.3-2m high, ascending from a procumbent base, with patent branches, subglabrous to puberulent; leaves ovate or lanceolate, 2.5-5cm long, irregularly dentate; petiole 0.25-4cm, auricles at base; capitula 0.75-2cm long, subhemispherical, solitary; peduncles 6-30cm; phyllaries linear-subulate; corollas yellow or orange, equalling or exceeding the pappus. Forest edge; 1600–1633m.
Distr.: SE Nigeria & W Cameroon to E Africa [afromontane].
IUCN: LC
Dom: Cheek 13513 fl., 9/2006; Dom photographic record 6 2/2006.

Dichrocephala integrifolia (L.f.) Kuntze subsp. *integrifolia*
Annual herb, 0.1-1m, erect or procumbent; leaves divided with a large terminal lobe and 1–3 pairs of smaller basal lobes, all dentate-serrate; corollas white, capitula 2–5mm long, in compound panicles; pappus absent or of few bristles. Forest & forest-grassland transition; 1636m.
Distr.: palaeotropical [montane].
IUCN: LC
Dom: Cheek 13524 fl., 9/2006.

Echinops gracilis O.Hoffm.
Spiny herb, 0.6-0.9(-2)m tall; cauline leaves dense, linear, stiff, 18 × 0.5-1mm, apex spinulose, margin with pinnately

patent spines c. 2mm long; capitula eligulate, mauve, 5cm diam. Grassland; 1931m.
Distr.: N Nigeria, Cameroon, Sudan, Congo (Brazzaville), Uganda [lower Guinea & Congolian (montane)].
IUCN: LC
Dom: Cheek sight record 25 9/2006.

Elephantopus mollis Kunth
Perennial herb, 40–90cm; leaves oblanceolate, 6–16 × 2–5cm, the base clasping; capitula aggregate into glomerules, these in compound cymes; capitula single-flowered; corollas white; pappus of 5–6 setae. Grassland, forest edge; 1633m.
Distr.: tropical America, introduced in Africa & Asia [pantropical].
IUCN: LC
Dom: Cheek 13500 9/2006.

Gynura pseudochina (L.) DC.
Prodr. 6: 299 (1838).
Syn. *Gynura miniata* Welw.
Perennial herb, 0.4-1m, unpleasantly aromatic; rootstock tuberous; leaves in rosette or leafy to mid-point, pubescent or glabrous, green or purplish, basal leaves ovate or spathulate, 4-22 × 2.5-11cm, margins entire, middle and upper leaves elliptic, obovate or narrowly obovate, lobed to pinnatisect, 6-25 × 2-7cm, margins lobed; capitula terminal; involucre campanulate, 8-13 × 13-15mm; corollas orange or yellow, 9-12.5mm; achenes 3mm long; pappus 7-11mm long. Grassland; 1490m.
Distr.: Sierra Leone to Ethiopia, southwards to Angola and Zambia, and tropical Asia [palaeotropics].
IUCN: LC
Dom: Harvey 325 2/2006.

Helichrysum forskahlii (J.F.Gmel.) Hilliard & B.L.Burtt
Syn. *Helichrysum cymosum* (L.) Less. subsp. *fruticosum* (Forssk.) Hedberg
Syn. *Helichrysum cymosum* sensu Adams
Syn. *Helichrysum helothamnus* Moeser var. *helothamnus* Bot. Jahrb. Syst. xliv: 259 (1910).
Perennial herb, or shrub, 0.2–1m high, densely leafy; leaves narrowly lanceolate, 0.4–3 × 0.1–1cm; capitula in clusters in corymbs; involucre and corollas pale yellow; pappus of short white bristles. Forest-grassland transition; 1780m.
Distr.: Nigeria to Zambia, Sudan to Tanzania & Yemen [afromontane].
IUCN: LC
Dom: Cheek 13544 fl., 9/2006.

Launaea nana (Bak.) Chiov.
Fl. Zamb. 6(1): 217 (1992); F.T.E.A. Compositae 1: 100 (2000).
Syn. *Sonchus elliotianus* Hiern
Perennial herb, with annual flowering shoots 1–90cm high from long fleshy taproot; leaves 2–6, in basal rosettes, usually appearing after flowering, (ob-)lanceolate to obovate, linear when young, to 25 × 7cm wide, margins entire to dentate to pinnatilobed; capitula few to many in corymbs; florets ligulate 9-18mm long, yellow; achenes with white setae 7-10mm long. Grassland; 1735m.
Distr.: W Africa, Congo (Kinshasa), Rwanda, Sudan, Kenya, Malawi, Mozambique, Zambia, Zimbabwe, Angola, South Africa [tropical Africa].
IUCN: LC
Dom: Dom photographic record 19 2/2006.

Mikaniopsis paniculata Milne-Redh.
Climbing, often woody herb; leaves alternate, long petiolate, lamina base cordate, margins minutely dentate, apices acuminate; inflorescences axillary, corymbose, many headed;

capitula short-stalked; involucres with calyculus of few lanceolate bracts; phyllaries c. 9; outer florets female, inner florets hermaphrodite; corollas whitish; pappus dirty white. 1831m.
Distr.: Cameroon, Nigeria & São Tomé [lower Guinea].
IUCN: NT
Dom: Cheek 13649 9/2006.

Solanecio mannii (Hook.f.) C.Jeffrey
Kew Bull. 41: 922 (1986); Fragm. Flor. Geobot. 36(1) Suppl.1: 418 (1991).
Syn. *Crassocephalum mannii* (Hook.f.) Milne-Redh.
Shrub, or soft-wooded tree, up to c. 7m high, subglabrous; leaves ovate, 10-20 × 2-5cm, acuminate, alternate, puberulous on midrib beneath; petiole 0.7-2.5cm; capitula oblong, c. 1cm long, on short slender pedicels in dense panicled cymes, c. 1cm long; phyllaries 5, linear, cymbiform; florets about 6, tubular; corollas yellow; achenes minutely hairy. Forest; 1735m.
Distr.: Nigeria to Congo (Kinshasa), Sudan to E Africa & Zimbabwe [afromontane].
IUCN: LC
Dom: Dom photographic record 7 2/2006.

Vernonia glabra (Steetz) Vatke
Perennial herb, with stems 0.3–3m from a woody rootstock; leaves elliptic, 2–16 × 0.5–7cm, serrate; capitula many in dense corymbose cymes; corollas blue; pappus of two rows of bristles, the outer smaller. Forest edge, thickets; 1490m.
Distr.: widespread in tropical Africa [tropical Africa].
IUCN: LC
Dom: Harvey 326 2/2006.

Vernonia myriantha Hook.f.
Syn. *Vernonia subuligera* O.Hoffm.
Syn. *Vernonia ampla* O.Hoffm.
Shrub or small tree, 1–6m; stems densely tomentose, hairs 'T-shaped'; leaves petiolate, upper sessile, elliptic or ovate, 11–50 × 2–23cm, serrate; inflorescences of thyrsoid cymes; phyllaries 3-6-seriate; florets few; corollas mauve or purple fading to white, scented; achenes ribbed; pappus biseriate, whitish. Forest; 1735m.
Distr.: Sierra Leone to Ethiopia & S Africa [afromontane].
IUCN: LC
Dom: Harvey 315 2/2006.

Vernonia purpurea Sch.Bip.
Perennial herb, to 2m, stems ridged, scabrid; leaves crowded, aromatic, sessile, elliptic or ovate to narrowly lanceolate, scabrid but denser beneath with glands; inflorescences of scorpioid cymes; phyllaries 4-8-seriate; often with inner recurved or spreading purple apices; florets many; corollas purple or mauve; achenes ribbed and pubescent; pappus biseriate, inner setae straw-coloured or violet. Montane grassland; 1633m.
Distr.: Senegal to Mozambique [afromontane].
IUCN: LC
Dom: Cheek 13494 fl., 9/2006.

Vernonia smithiana Less.
Perennial herb, to 1m from a woody stock; capitula c. 0.75cm wide, hemispherical; corollas pinkish or reddish-purple. Grassland; 1633m.
Distr.: W Africa to E Africa and south to Angola [tropical Africa].
IUCN: LC
Dom: Cheek 13517 9/2006.
Note: Cheek 13517 is sterile, but matches this taxon vegetatively (Cheek, xi.2008).

Connaraceae

M. Cheek (K)

Agelaea pentagyna (Lam.) Baill.
Agric. Univ. Wag. Papers 89(6): 144 (1989); Fl. Gabon 33: 34 (1992).
Syn. *Agelaea hirsuta* De Wild.
Syn. *Agelaea dewevrei* De Wild. & T.Durand
Syn. *Agelaea floccosa* Schellenb.
Syn. *Agelaea grisea* Schellenb.
Syn. *Agelaea obliqua* (P.Beauv.) Baill.
Syn. *Agelaea preussii* Gilg
Syn. *Agelaea pseudobliqua* Schellenb.
Liana, 6(–25)m; branches furrowed, puberulent; leaves trifoliolate, 14 × 14cm, leaflets ovate, terminal leaflet 10 × 8cm, acumen 1.5cm, base rounded, finely tomentose on lower surface and veins; petiole 11.5cm; petiolule 3mm; panicles to 35cm, glabrous to tomentose; sepals 2.5–5mm, with fringing multi-cellular hair; petals 3–5.5mm. Forest; 1600m.
Distr.: Senegal to Mozambique & Madagascar [tropical Africa & Madagascar].
IUCN: LC
Dom: Cheek 13444 9/2006.
Note: *Agelaea pentagyna* s.l., numerous taxa appear to be sunk under this name (Cheek, xi.2008).

Convolvulaceae

Y.B. Harvey (K)

Ipomoea involucrata P.Beauv.
F.T.E.A. Convolvulaceae: 104 (1963); F.W.T.A. 2: 347 (1963); Fl. Zamb. 8(1): 75 (1987);
Climber, densely velutinous hairs; sheathing involucral bract entire, c. 1.5 × 7cm; flowers 5–15, capitate; corolla pink, 3–4cm. Farmbush, grassland and woodland weed; 1490m.
Distr: W Africa to Angola and South Africa [tropical Africa].
IUCN: LC
Dom: Dom photographic record 37 2/2006;
Note: very common weed, identified by the sheathing involucral bract.

Crassulaceae

E. Ndive (SCA) & B. Nke (YA)

Crassula alsinoides (Hook.f.) Engl.
Fleshy straggling herb; stems with 2 lines of pale hairs; leaves ovate, 1–1.5 × 0.5–1cm, apex acute, base claw-like then short-attenuate; flowers terminal or axillary; pedicel fine, to 7mm; flowers 0.5cm; sepals lanceolate; petals lanceolate, white. Forest edge, open areas; 1636m.
Distr.: W Cameroon & Bioko to E & S Africa, Yemen [afromontane].
IUCN: LC
Dom: Cheek 13521 fl., 9/2006.

Crassula schimperi Fisch. & Mey. subsp. *schimperi*
F.T.E.A. Crassulaceae: 7 (1987).
Syn. *Crassula pentandra* (Edgew.) Schönl.
Small herb, erect, 5-10cm; leaves lanceolate, c. 7mm long, broadened towards the base; flowers sessile. Forest, sometimes on rocks; 1931m.

Distr.: Mount Cameroon, Congo (Kinshasa), Sudan, Ethiopia, Zambia, E Africa, tropical Arabia and India [montane].
IUCN: LC
Dom: Cheek 13588 fl., 9/2006.

Crassula vaginata Eckl. & Zeyh.
F.T.E.A. Crassulaceae: 13 (1987).
Syn. *Crassula alba* sensu Hepper
Syn. *Crassula mannii* Hook.f.
Erect herb, to 90cm; stem stout; leaves linear-lanceolate, to 10cm long; flowers white. Grassland and forest-grassland transition; 1797m.
Distr.: W Cameroon to E and South Africa [afromontane].
IUCN: LC
Dom: Cheek 13568 fl., 9/2006.

Kalanchoe crenata (Andrews) Haw.
Syn. *Kalanchoe laciniata* sensu Hepper
Erect, unbranched, succulent herb, to 1m; leaves elliptic, c. 7 × 4.5cm, rounded, obtuse, crenate; petiole 1cm; inflorescence corymbose, 6cm across, c. 20-flowered; pedicels glandular-pubescent; flowers pink, yellow or red, 1.5cm long; petals 5. Farmbush; 1600m.
Distr.: Guinea (Conakry) to Cameroon [upper & lower Guinea].
IUCN: LC
Dom: Onana 3616 2/2006.

Cucurbitaceae

L. Pearce (K)

Fl. Cameroun 6 (1967).

Coccinia barteri (Hook.f.) Keay
Herbaceous climber; stems ridged, glabrous, angular; tendrils bifid; leaves variable, palmatisect (3–)5 lobed to sub-lobed, 12–20 × 10–15cm, base ± deeply cordate, lobe apices acute or acuminate, margins ± serrate; dioecious, both sexes on racemes; peduncle 2–3cm, pubescent; flowers subsessile; males with 5–10 flowers, females 3–5; flowers c. 1.8 × 1.5cm; petals yellow; fruits ellipsoid or suborbicular c. 4 × 2.5cm, green, drying blackish, smooth; seeds numerous, c. 4 × 2mm, white, smooth with narrow rim. Farmbush & forest; 1659m.
Distr.: widespread in (sub-)tropical Africa [tropical Africa].
IUCN: LC
Dom: Cheek 13536 9/2006.
Uses: MEDICINES (Cheek 13536).

Momordica foetida Schum. & Thonn.
Syn. *Momordica cordata* Cogn.
Herbaceous climber; stems deeply ridged, finely pubescent or glabrous; leaves ovate, 9–14 × 8–12cm, base deeply cordate, apex acuminate, margin shallowly dentate, pubescent along veins on abaxial lamina; petioles c. 7cm; monoecious, male flowers in umbels; peduncle c. 10cm, finely pubescent, may be kinked; pedicels 1–5cm subtended by a leafy bract c. 4mm; sepals rounded at apex, pubescent, dry black; petals obovate 1.5 × 0.5cm; stamens 2 with sinuous anthers; female flowers solitary; pedicel c. 7cm; fruits densely hispid when immature, pale orange with sparser bristling when mature, ovoid to ellipsoid c. 5.5 × 3cm, seeds red. Secondary forest & farmbush; 1600m.
Distr.: tropical & S Africa [tropical & subtropical Africa].
IUCN: LC
Dom: Cheek 13405 fl., 9/2006; Nana 138 4/2005.

Zehneria scabra (L.F.) Sond.

F.T.E.A. Cucurbitaceae: 122 (1967); Fl. Cameroun 6: 44 (1967).

Syn. *Melothria punctata* (Thunb.) Cogn.
Syn. *Melothria mannii* Cogn.

Herbaceous climber or trailer; stems 1–2mm, ridged, scabrid around nodes; leaves 5–6(–9) × 3.5–4(–7.5)cm, ovate, base deeply cordate, apex acute or acuminate, margin dentate, scabrous, particularly on upper surface, drying blackish; dioecious; male umbels c. 20-flowered; peduncle 1–3cm, finely pubescent; pedicels 0.5cm; female umbels 5(–10)-flowered; peduncle 1–2mm; flowers 1–2mm; calyx pubescent, blackish; petals white-yellow; fruit orbicular, 0.5–1cm diam., orange when mature, finely pitted. Secondary vegetation, farmland; 1700–1800m.
Distr.: tropical Africa & Asia [palaeotropical].
IUCN: LC
Dom: Corcoran 12 4/2005; Nana 155 4/2005.

Ericaceae

J.-M. Onana (YA) & M. Cheek (K)

Fl. Cameroun 11 (1970).

Agarista salicifolia G.Don

Gen. Syst. 3: 837 (1834).

Syn. *Agauria salicifolia* (Comm. ex Lam.) Hook.f. ex Oliv.

Shrub, or small tree, to 13m; leaves oblong-lanceolate, acute at ends; inflorescence a short axillary raceme; flowers pink. Forest edge; 1633m.
Distr.: Cameroon to Madagascar [afromontane & Madagascar].
IUCN: LC
Dom: Cheek 13504 fl., fr., 9/2006.

Erica mannii (Hook.f.) Beentje

Utafiti 3: 13 (1990).

Syn. *Philippia mannii* (Hook.f.) Alm & Fries

Heath-like shrub, to 4m; leaves ascending to erect, in whorls of 3; few purplish flowers at end of each branchlet. Forest edge; 1780m.
Distr.: Cameroon to Kenya [afromontane].
IUCN: LC
Dom: Cheek 13541 9/2006.

Erica tenuipilosa (Engl. ex Alm & T.C.E.Fr.) Cheek subsp. tenuipilosa

Kew Bull. 52: 753 (1997).

Syn. *Blaeria mannii* (Engl.) Engl.
Syn. *Blaeria spicata* Hochst. ex A.Rich. subsp. *mannii* (Engl.) Wickens
Kew Bull. 27: 513 (1972).
Syn. *Blaeria spicata* Hochst. ex A.Rich. var. *mannii* (Engl.) Letouzey
Fl. Cameroun 11: 196 (1970).
Syn. *Blaeria spicata* Hochst. ex A.Rich. var. *fakoensis* Letouzey
Fl. Cameroun 11: 195 (1970).
Syn. *Blaeria spicata* Hochst. ex A.Rich. var. *nimbana* (A.Chev.) Letouzey
Fl. Cameroun 11: 198 (1970).

Heath-like undershrub, 0.3-0.8m; leaves in whorls of 3; inflorescence a spicate panicle; flowers purplish. Forest-grassland transition; 1780m.
Distr.: Ivory Coast to Bioko and W Cameroon [Guineo-Congolian (montane)].
IUCN: LC
Dom: Cheek 13542 9/2006.

Note: Cheek 13542 is a field determination and has yet to be verified (Cheek, vi.2009).

Euphorbiaceae

B. Tchiengue (YA), B. Oben (K), M. Cheek (K), O. Sene (YA), E. Ndive (SCA), B. Nke (YA) & S. Williams

Antidesma pachybotryum Pax & K.Hoffm.

Enum. Pl. Afr. Trop. 1: 206 (1991).

Tree or shrub, 4-5m; long brown hairy; leaves oblong or obovate-oblong, 18-21 × 7-9.5cm, shortly acuminate, base rounded or truncate, 6-8 lateral nerve pairs; petioles 0.8-1.2cm; stipules triangular, 8mm; inflorescences 7-14cm; fruits red. Forest, often near streams; 1633m.
Distr.: Cameroon Highlands; Cameroon, CAR [W Cameroon Uplands].
IUCN: VU
Dom: Cheek 13469 9/2006.

Antidesma venosum Tul.

Tree, 6m, puberulent; leaves obovate, 10-15 × 4- 6.5cm, apex rounded or truncate, base obtuse; stipules entire; infructescence pendent, 10cm; fruit 6mm, ovate, flattened, juicy. Farms; 1600m.
Distr.: tropical Africa [tropical Africa].
IUCN: LC
Dom: Cheek 13418 fr., 9/2006.
Local name: Fijeh-Fa (Cheek 13418).

Antidesma vogelianum Müll.Arg.

Tree or shrub, 3-10m, pubescent; leaves elliptic, rarely slightly obovate, c. 15 × 5.5cm, acumen c. 2cm, base acute; petiole c. 5mm; stipule lanceolate, c. 6 × 2mm, entire; infructescence non-interrupted, pendulous, racemose, c. 15cm; fruits ellipsoid, 7mm, red, fleshy. Forest; 1640m.
Distr.: Nigeria to Tanzania [lower Guinea & Congolian].
IUCN: LC
Dom: Pollard 1383 4/2005.

Bridelia micrantha (Hochst.) Baill.

Shrub or tree, 2-14m; trunk spiny, glabrescent; leaves oblong-elliptic, 10-15 × 3-5cm, acuminate, drying pale green below (bluish green live), the secondary nerves fawn, joining to form a marginal nerve; inflorescence axillary glomerules; flowers greenish white, 3mm diam.; fruit axillary, sessile, fleshy, ellipsoid, red, 7mm. Forest edge; 1600–1735m.
Distr.: Senegal to S Africa [tropical Africa].
IUCN: LC
Dom: Cheek 13421 fr., 9/2006; Harvey 313 2/2006; Pollard 1393 4/2005.

Croton macrostachyus Hochst. ex Del.

Tree, 6m; leaves densely stellate hairy above, broadly ovate, c. 12–8.5cm, acute, deeply cordate, margin obscurely hooked, basal glands sessile; inflorescence 10cm; fruit 1.2cm diam. Roadside, forest edges, cultivated; 1640m.
Distr.: tropical Africa [tropical Africa].
IUCN: LC
Dom: Nana 146 4/2005.

Erythrococca membranacea (Müll.Arg.) Prain

Shrub, 2–3(–5)m; unarmed, pilose, hairs yellowish brown, c. 2mm long; leaves oblong or elliptic, 8–12 × 3.5–5cm, long acuminate, obtuse, margin coarsely and irregularly serrate; inflorescence 6–8-flowered, subumbellate, 3cm; peduncle 2cm; pedicels 5mm; flowers green, 3mm wide; fruit green, bilobed, 1.5cm wide, seeds red. Forest; 1450m.
Distr.: SE Nigeria & W Cameroon [lower Guinea].

IUCN: NT
Dom: <u>Harvey 321</u> 2/2006.

Erythrococca sp. aff. *anomala* (Juss. ex Poir.) Prain

Shrub, to 2m; stems lacking spines, with minute fingernail-like brown scales at nodes, long, appressed hairy; leaves elliptic, 4-7 × 1.5-3cm, long, acuminate, sparsely serrate; inflorescence sessile, sparsely flowered; female flowers green, globose, 2mm; stigmas laciniate; fruit biglobose, 7mm wide, green, dehiscent. Secondary forest; 1640m.
Dom: <u>Pollard 1386</u> 4/2005.
Note: Pollard 1386 lacks the spines of *E. anomala* but has the laciniate stigmas. Hybrid with *E. hispida* or new taxon? (Cheek, vi.2009).

Macaranga occidentalis (Müll.Arg.) Müll.Arg.

Tree, to 3–25m; trunk often spiny, with red or clear exudate, glabrous; leaves suborbicular, 30–50cm diam., shallowly 3–5-lobed, lobes acuminate, base cordate, lower surface often bluish white below and with minute red glands; petiole c. 30cm; stipules 4 × 2cm; inflorescence axillary, pendent, paniculate; bracts 5–10cm, deeply dentate-sinuate, densely pubescent; female inflorescences to 18cm; fruit 1–2-lobed, 8mm. Forest edge; 1600m.
Distr.: SE Nigeria & Cameroon [Western Cameroon Uplands].
IUCN: NT
Dom: <u>Onana 3618</u> 2/2006.

Margaritaria discoidea (Baill.) G.L.Webster var. *discoidea*

J. Arnold. Arb. 48: 311 (1967); Meded. Land. Wag. 82(3): 145 (1982); F.T.E.A. Euphorbiaceae (2): 63 (1988); Keay R.W.J., Trees of Nigeria: 167 (1989); Ann. Missouri Bot. Gard. 77: 217 (1990); Hawthorne W., F.G.F.T. Ghana: 85 (1990).
Syn. *Phyllanthus discoideus* (Baill.) Müll.Arg.
Deciduous tree, 10–25m, glabrous; leaves ovate–elliptic, 7–9 × 3.5cm, subacuminate, acute to obtuse, bluish white below, drying green, secondary nerves 10 pairs, arcuate, quaternary nerves reticulate, conspicuous; petiole 8mm; fruit 3–lobed, glossy, 7mm. Farmbush & forest edge; 1735m.
Distr.: widepsread in tropical Africa [tropical Africa.].
IUCN: LC
Dom: <u>Dom photographic record 28</u> 2/2006.

Pedilanthus tithymaloides (Linn.) A.Poit.

Shrub, 1.5m; white exudate; stems fleshy, zig-zag; leaves elliptic, to 8 × 3cm long, vareigated green & white; flowers 1-2, terminal, zygomorphic, red, 2cm. Cultivated; 1797m.
Distr.: C & S America, introduced in Africa & Asia [tropical America & cultivated].
IUCN: LC
Dom: <u>Cheek 13668</u> fl., 9/2006.
Uses: ENVIRONMENTAL USES – Boundaries/Barriers/Supports - cultivated as a living hedge; MEDICINES (Cheek 13668).

Ricinus communis L.

Annual herb, to 2m; leaves orbicular, to 60cm broad, deeply palmately lobed, green or reddish; inflorescence a large panicle; male flowers below, females above; capsule smooth or prickly, 2.5cm diam. Cultivated; 1831m.
Distr.: throughout Africa, and cultivated and naturalized throughout the tropical, subtropical and warm temperate regions of the globe [tropical Africa & cultivated].
IUCN: LC
Dom: <u>Cheek sight record 42</u> 9/2006.

Shirakiopsis elliptica (Hochst.) H.-J.Esser

Kew Bull. 56(4): 1018 (2001).
Syn. *Sapium ellipticum* (Krauss) Pax
Deciduous shrub or tree, 3–25m; white exudate, glabrous; leaves drying black, leathery, elliptic, c. 9 × 4cm, obtuse, acute, finely serrate; petiole 3mm; inflorescence terminal spike, 6cm; flowers numerous, 1mm, green; fruit globose or bilobed, fleshy, 8mm; styles 2, coiled. Forest or forest edge; 1640m.
Distr.: tropical & S Africa [tropical Africa].
IUCN: LC
Dom: <u>Onana 3635</u> 2/2006.

Gentianaceae

M. Cheek (K)

Sebaea brachyphylla Griseb.

Erect herb, 20(–50)cm; stems glabrous, few-branched, dichotomous; leaves sessile, orbicular, 8mm; cymes terminal, dense, 5–20-flowered; pedicels c. 1mm; sepals oblong, 3mm with a prominent central nerve; corolla 6mm, inflated at base, yellow; stamens exserted. Grassland; 1931m.
Distr.: tropical Africa [afromontane].
IUCN: LC
Dom: <u>Cheek 13572</u> fl., 9/2006.

Swertia eminii Engl.

Enum. Pl. Afr. Trop. 4: 355 (1997).
Erect herb, to 25cm, glabrous; lateral branches few, c. 3cm; leaves subsessile, elliptic-lanceolate, c. 13 × 4.5mm, strongly 3(–5) parallel-nerved; cymes terminal, 3–6-flowered; pedicels 6mm; sepals ovate-lanceolate, 4–5mm; petals 4–5, ovate, 5–7mm, white. Grassland; 1633–1931m.
Distr.: Cameroon, Congo (Kinshasa), Uganda, Burundi, Sudan, Kenya [afromontane].
IUCN: LC
Dom: <u>Cheek 13511</u> fl., 9/2006; <u>13607</u> fl., 9/2006; <u>13626</u> fl., 9/2006.
Note: Gentians of Cameroon desperately need revision. The determination for Cheek 13607 is likely to need changing (Cheek, xi.2008). Cheek 13511 & 13626 are 4-merous, not 5.

Swertia mannii Hook.f.

Erect herb, to 15cm; lateral branches to 8cm; leaves linear-lanceolate to narrowly elliptic to 1-3cm long; cymes more lax, 6–8-flowered; pedicels to 1.8cm; sepals lanceolate, 3mm; petals 5, lanceolate, 5.5mm, white with a purple stripe externally. Grassland; 1797m.
Distr.: Guinea (Conakry) to Cameroon [upper & lower Guinea].
IUCN: LC
Dom: <u>Cheek 13562</u> fl., 9/2006.
Note: Cheek 13562 has slender, purple leaves etc., this taxon really needs revising (Cheek, xi.2008).

Geraniaceae

J.-M. Onana (YA)

Geranium arabicum Forssk. subsp. *arabicum*

Notes Roy. Bot. Gard. Edinburgh 42: 171 (1985).
Syn. *Geranium simense* Hochst. ex A.Rich.
Straggling herb, 6–20cm; leaves orbicular in outline, 2–3cm diam., deeply palmately lobed, lobes 5, with 2–3 lateral lobes, sparingly pubescent; petiole 3-6cm; inflorescence 1–2-flowered; peduncle c. 8cm; flowers 1cm, pale pink (-white) with dark veins. Grassland-forest boundary, roadsides; 1797m.
Distr.: Nigeria to Kenya [afromontane].

IUCN: LC
Dom: Cheek 13565 fl., 9/2006.
Note: Cheek 13565 is the only record of this taxon at Dom.
The identification was made in the field and has yet to be
verified (Harvey, iii.2009).

Guttiferae

M. Cheek (K) & J.-M. Onana (YA)

Garcinia conrauana Engl.
Tree, (4–)15(–20)m, bole cylindrical; leaves drying pale
brown below, elliptic c. 14 × 10cm, apex rounded, acumen
abrupt, 0.5cm, base obtuse, margin revolute, resin canals
concolorous; petiole 2.5cm; inflorescence axillary, subsessile,
3-flowered; flowers white, 3cm diam., staminal mass of four
segments joining to form a dome 1.5 × 1cm; fruit spherical,
green, 12 × 12cm. Submontane forest; 1980m.
Distr.: W Cameroon [lower Guinea].
IUCN: NT
Dom: Pollard 1398 4/2005.
Local name: Iroko (Pollard 1398).

Garcinia smeathmannii (Planch. & Triana) Oliv.
Syn. *Garcinia polyantha* Oliv.
Tree, 4–15m; leaves thickly leathery, drying pale brown
below, narrowly oblong-elliptic, c. 20 × 8cm, obtuse,
subacuminate, base obtuse, secondary nerves c. 20 pairs;
petiole c. 1.5cm; inflorescence sessile, axillary, on leafy
stems, umbellate-fascicled, 15–20-flowered; pedicels c.
15mm; flowers white, 1cm diam.; anthers with free filaments
inserted on ligules, as long as petals; fruits globose, 2cm;
stigmas 2; pedicels 4cm. Forest; 1570–1980m.
Distr.: Guinea (Bissau) to Zambia [Guineo-Congolian].
IUCN: LC
Dom: Cheek 13416 9/2006; 13622 fr., 9/2006; Nana 151
4/2005; Onana 3603 2/2006; 3633 2/2006; Pollard 1395
4/2005.
Local name: Teleh (Cheek 13416).
Note: Onana 3603 has unusually short stalked fruits (MC,
ix.2008).

Harungana madagascariensis Lam. ex Poir.
Shrub or small tree, 3–6m, glabrescent; leaves ovate c. 12 ×
5.5cm, acuminate, base rounded, nerves c. 10 pairs; petiole
2cm, producing bright orange exudate when broken;
inflorescence a dense terminal panicle, 7–15cm; flowers
white, 2mm, petals hairy; berries orange, 3mm. Farmbush;
1450–1931m.
Distr.: tropical Africa & Madagascar [tropical Africa].
IUCN: LC
Dom: Cheek sight record 26 9/2006; Corcoran sight record 6
2/2006.

Hypericum revolutum Vahl subsp. *revolutum*
Webbia 22: 239 (1967); Bull. Jard. Bot. Nat. Belg. 41: 438
(1971); Kew Bull. 33: 581 (1979).
Syn. *Hypericum lanceolatum* Lam.
Shrub or tree, to 12m; leaves narrowly elliptic, 1–2(–3.5)cm
long; petals yellow, 2–3cm long. Grassland & forest-
grassland transition; 1931m.
Distr.: tropical Africa [afromontane].
IUCN: LC
Dom: Cheek 13576 9/2006.

Psorospermum aurantiacum Engl.
Shrub or small tree, to 2(–5)m; young stems rusty-tomentose;
leaves bullate, elliptic, 3.5–7 × 1.8–3.5cm, apex shortly
acuminate, base acute, brown-green and sparsely tomentose
above, venation impressed, densely rusty-tomentose below,

obscuring the venation; petiole 4–5mm, tomentose;
inflorescence terminal on lateral branches, pubescent
throughout, 10–many-flowered; sepals acute, 2.5–3mm;
petals cream, 4.5mm, pubescent within; fruit ovoid, wine-red.
Grassland and forest edge; 1600–1633m.
Distr.: Nigeria and Cameroon [Western Cameroon Uplands].
IUCN: VU
Dom: Cheek 13404 9/2006; 13516 9/2006; Onana 3627
2/2006.

Psorospermum densipunctatum Engl.
Shrub or small tree, to 2(–3)m; young stems sparsely
pubescent; leaves densely bullate, elliptic, 3–5 × 1.8–2.8cm,
apex shortly acuminate, base acute, upper surface dark green,
glossy, venation deeply impressed, lower surface pubescent
only on midrib and ± on lateral nerves; petiole 3–6mm, ±
pubescent; inflorescence terminal on lateral branches, cymes
subumbellate, c. 10–20-flowered, puberulent throughout;
sepals acute, c. 2.5mm; petals white to cream, pubescent
within; fruit ovoid, wine-red. Grassland and forest edge;
1600m.
Distr.: Sierra Leone, Nigeria, Cameroon [upper & lower
Guinea (montane)].
IUCN: NT
Dom: Onana 3617 2/2006.
Note: Onana 3617 has black glands atypically hard to see on
the lower leaf surface (Cheek, ix.2008).

Psorospermum cf. *tenuifolium* Hook.f.
Shrub or climber, to 4m; lateral shoots puberulent; leaves in
2–3 pairs on lateral branches, opposite or rarely sub-opposite,
papery, concolorous, elliptic-obovate, c. 6.5 × 3.4cm, apex
shortly acuminate, base cuneate, margin sub-crenulate, lower
surface densely punctate; petiole c. 5mm, puberulent; cymes
terminal on lateral branches, umbellate, 8–30-flowered;
peduncle 1.7cm; pedicels 0.4cm, puberulous; sepals acute, c.
2mm long, puberulent; petals elliptic, c. 4mm long, white,
dark-veined, densely white-pubescent within. Forest; 1490m.
Distr.: [W Cameroon Uplands].
Dom: Harvey 328 2/2006.
Note: Harvey 328 matches the description under this name of
Bali Ngemba, although it is unusually hairy (Cheek, ix.2008

Symphonia globulifera L.f.
Tree, to 30m, glabrous; leaves narrowly elliptic to oblong-
elliptic, c. 10.5 × 2.8cm, long acuminate, acute, lateral nerves
c. 30 pairs, resin canals inconspicuous; petiole 0.7cm;
inflorescence terminal, c. 10-flowered; pedicels 1.5cm;
flowers red, globose, c. 1cm diam.; styles ascending,
aculeate, 5; fruit ovoid, 2.5cm, lenticellate; styles persistent.
Forest; 1633–1831m.
Distr.: tropical America, Africa & Madagascar [amphi-
Atlantic].
IUCN: LC
Dom: Cheek 13481 9/2006; 13645 9/2006.

Icacinaceae

M. Cheek (K)

Fl. Cameroun 15 (1973).

Rhaphiostylis beninensis (Hook.f. ex Planch.)
Planch. ex Benth.
Woody climber, to 10–12m; leaves elliptic to lanceolate-
elliptic, 6–15 × 2–7cm, 5–7 pairs lateral nerves; flowers
rather numerous in axillary facicles, pentamerous; petals free,
7 × 1mm, drying black; stamens 5, white; pedicels 7–8mm;
fruits red, reniform, 1.3 × 1–2 × 1cm, reticulate. Forest;
1633m.

Distr.: Senegal to Tanzania [Guineo-Congolian].
IUCN: LC
Dom: Cheek 13466 9/2006.
Note: Cheek 13466 does not match venation of fertile material well, but is closer to plot vouchers from Guinea where it is common near this altitude. Venation matches *Ptychopetalum petiolatum* better but petioles and stems wrong (Cheek, x.2008).

Labiatae

A. Paton (K), G. Bramley (K), M. Cheek (K) & B.J. Pollard

Achyrospermum aethiopicum Welw.
F.T.A. 5: 465 (1900).
Single-stemmed shrubby herb, 0.5-2m tall; stems pubescent to pilose, gradually glabrescent; leaves petiolate, blade elliptic to elliptic-ovate, c. 10-20 × 4-8cm, coarsely serrate, acuminate, base cuneate and attenuate, subglabrous; petiole 2-8cm; inflorescence terminal, on very short axillary branches, subsessile to shortly pedunculate; small sterile bracts, 10-70, subtending 3-5 flowers, elliptic-ovate, 3-10mm long; bracteoles linear-elliptic, c. 1mm long; pedicels 1-2mm long; calyx 6-8mm long, pubescent; corolla white to lilac or pink, 6-8mm long, posterior lip 1-1.2mm long, emarginate, anterior lip 2-2.5mm long, lateral lobes 0.6-0.8mm long; nutlets dull brown, oblong-ellipsoid, c. 2.5 × 1.3-1.4mm. Riverine forest; 1570–1831m.
Distr.: Cameroon, Congo (Kinshasa), Angola, Malawi, Zambia [tropical Africa].
IUCN: LC
Dom: Cheek 13638 fl., 9/2006; Onana 3630 2/2006.

Aeollanthus repens Oliv.
Erect perennial herb, 15-30cm, sharp scented when crushed; tuber not seen, hairs white, patent, sparse, 0.5-1mm; leaves linear, c. 30 × 2mm; inflorescences spike-like, 10 × 5mm; flowers bright purple blue, 1.5cm, numerous. Rocky outcrops; 1797m.
Distr.: Cameroon, Sudan, Uganda, Kenya, Tanzania [montane].
IUCN: LC
Dom: Cheek 13550 fl., 9/2006.
Note: common but very local.

Clerodendrum carnosulum Baker
Climber, stems square, 5mm wide, minutely hairy when young, spines absent; leaves of non-flowering stems ovate, to 14 × 7.5cm, acute-acuminate, base truncate to shallowly cordate, lateral nerves 3-4 pairs, minutely hairy below; petiole 3cm; leaves of flowering stems elliptic, to 6 × 3cm; petiole 1cm; inflorescence terminal on leafy spur shoots 6-10cm long, 8-10-flowered; calyx red, saucer-shaped, 5 × 10mm, shallowly 5-lobed; corolla blue or purple, 15mm long, strongly bilateral. Grassland, forest edges; 1490m.
Distr.: Cameroon Highlands and Gabon (possibly Congo (Kinshasa) also) [lower Guinea].
IUCN: NT
Dom: Harvey 322 2/2006.

Clerodendrum silvanum Henriq. var. *buchholzii* (Gürke) Verdc.
Mem. Mus. Natl. Hist. Nat. B. Bot. 25: 1555 (1975); F.T.E.A. Verbenaceae: 110 (1992).
Syn. *Clerodendrum buchholzii* Gürke
Syn. *Clerodendrum thonneri* Gürke
Woody climber, to 10m; stems with petiolar spines; leaves elliptic or ovate, glabrous, 8–20 × 3–10cm, often confined to the canopy; inflorescence an elongate, leafless panicle, frequently cauliflorous; rachis 5–30cm long; calyx enlarged, 8–10mm; corolla white, fragrant; tube (1.5–)1.7–2.5cm; fruits red. Forest; 1570–1600m.
Distr.: tropical & subtropical Africa [tropical & subtropical Africa].
IUCN: LC
Dom: Cheek 13456 fl., 9/2006; Onana 3631 2/2006.
Note: Cheek 13456 is immature (Pollard, xi.2008).

Clerodendrum violaceum Gürke
Straggling or climbing shrub; young shoots quadrangular; leaves distinctly petiolate, thinly membranaceous, ovate, elliptic or oblong, 5–12 × 3–10cm; inflorescence paniculate; calyx lobes obtuse; flowers about 2.5cm, a conspicuous violet, violet and white or greenish. Forest; 1550m.
Distr.: Guinea (Conakry) to Cameroon, Congo (Kinshasa), Zimbabwe [tropical Africa].
IUCN: LC
Dom: Pollard 1379 4/2005.

Clerodendrum sp. 1 of Dom
Climber; glabrous; petiolar spines; leaf blades ovate, to 9 × 5.5cm, acumen 1.5cm, acute, with terminal mucro, base obtuse, margin entire, trinerved, lateral nerves arising from midrib 0.5cm from base, midrib deeply sunk above, with 3 pairs of weaker lateral nerves, looping 0.5cm from margin; petioles 1cm; flowers and fruits unknown. Submontane forest patches; 1831m.
Dom: Cheek 13627 9/2006.
Note: Cheek 13627 not matching any of the other three *Clerodendrum* species from Dom. Perhaps closest to *C. violaceum* but this is not trinerved. Flowering or fruiting material needed to identify (Cheek, vi.2009).

Leucas deflexa Hook.f.
Straggling or semi-erect aromatic herb, to 2m; leaves lanceolate, cuneate at base, serrate, with an entire, acute tip, petiolate; inflorescence a densely globose axillary whorl with numerous linear-subulate bracteoles; corolla white; stamens ascending; anthers often conspicuously hairy, orange. Forest, forest margins, savanna.
Distr.: Ghana, Bioko, Cameroon, Angola [Guineo-Congolian (montane)].
IUCN: LC
Dom: Pollard sight record 10 4/2005.

Leucas oligocephala Hook. f.
Herb, to 1.2m; stems slender, branched, pilose; leaves linear-lanceolate to elliptic, 2.5 or more × 1cm; inflorescence of several verticils of dense, globose, axillary whorls; calyx with longest tooth on the lower side, densely ciliate; flowers densely hairy and purple tinged in bud, white when open; stamens ascending; anthers orange or red. Grassland and forest-grassland transition; 1633m.
Distr.: W Africa to E Africa and South Africa [afromontane].
IUCN: LC
Dom: Cheek 13499 fl., 9/2006.
Note: Sebald's varieties are not recognised here because there are major overlaps between his distinguishing characters.

Micromeria imbricata C.Chr.
Dansk. Bot. Ark. iv. No. 3: 21 (1922).
Syn. *Satureja punctata* (Benth.) Briq.
A robust narrow erect or spreading heath-like woody herb, to ± 60cm; stems wiry, branched, pubescent; leaves ovate, 4 – 10mm long, conspicuously glandular punctate beneath, subsessile, entire; flowers few together, in axillary whorls; calyx tubular, 4mm long; corolla rich pink. Grassland and forest-grassland transition; 1633m.
Distr.: Cameroon to E and South Africa [afromontane].

IUCN: LC

Dom: Dom photographic record 16 2/2006; Cheek sight record 13 9/2006.

Ocimum gratissimum L. subsp. *gratissimum* var. *gratissimum*

A branched erect pubescent shrub, to ± 3m, emanating an aroma similar to that of cloves (*Syzygium aromaticum*); leaves ovate to obovate, 6–12 × 3cm, cuneate, acutely acuminate; inflorescence of several dense spikes, >1cm; calyx dull, densely lanate, horizontal or slightly downward pointing in fruit; corolla small, greenish-white; stamens declinate. Submontane forest, woodland, savanna; 1600m.

Distr.: widespread in the tropics from India to W Africa, S to Namibia & S Africa; naturalised in tropical S America [pantropical].

IUCN: LC

Dom: Cheek 13410 fr., 9/2006; Onana 3626 2/2006.

Local name: Munjiy (Cheek 13410).

Ocimum lamiifolium Hochst. ex Benth.

Perennial herb or shrub, to 3m; stems much-branched, woody; leaves ovate, 15-65 × 6-40mm, serrate, sparsely pubescent; petiole 3-14(-20)mm; inflorescence lax, verticils 1-2cm distant; calyx 3-5mm at anthesis, 8-11mm in fruit, often purplish; corolla white, white marked pink, or dull mauve, 9-11mm; stamens exserted 3-4mm, declinate. Scrubby forest, clearings, wasteland, secondary bush, forest edges; 1659m.

Distr.: W Cameroon to E Africa, Zambia, Malawi [afromontane].

IUCN: LC

Dom: Cheek 13528 fl., 9/2006.

Local name: Lui Lui (Cheek 13528). **Uses:** MEDICINES (Cheek 13528).

Platostoma denticulatum Robyns

Bull. Jard. Bot. Brux. 17: 22 (1943).

Perennial herb, 15-20cm; stems ascending; leaves ovate or trullate; blades 3-4 × 2-3cm, acute, base decurrent, margin with 4-5 deep teeth, glabrous; inflorescence erect, slender, 7-10cm, verticils 1cm, distinct; calyx 2mm, lower lip denticulate; corolla purple, 1-2mm. Damp places; 1600m.

Distr.: Cameroon, Congo (Kinshasa), Rwanda, Burundi, Uganda, Kenya, Tanzania, Angola [afromontane].

IUCN: LC

Dom: Cheek 13423 fl., 9/2006.

Uses: FOOD ADDITIVES – Unspecified parts – used in cooking (Cheek 13423).

Platostoma rotundifolium (Briq.) A.J.Paton

Kew Bull. 52 (2): 257 – 292. (1997).

Syn. *Geniosporum rotundifolium* Briq.

A stout woody perennial, to ± 2m; stems grooved, densely ferrugineous-pubescent; leaves subrotund to broadly lanceolate, 2–5 × 1–3cm, crenulate inflorescences several, dense, cylindrical, 2.5–10cm long; bracts broadly ovate, conspicuously white or mauve-tinged; calyx tubular, c. 4mm long at maturity, 4-toothed; corolla twice as long as calyx. Forest edge, grassland; 1600m.

Distr.: Sierra Leone to Cameroon, Congo (Kinshasa), E Africa, Angola [afromontane].

IUCN: LC

Dom: Dom photographic record 18 2/2006.

Plectranthus bojeri (Benth.) Hedge

F.T.E.A. Lamiaceae: 327 (2009).

Syn. *Solenostemon latifolius* (Hochst. ex Benth.) J.K.Morton

Decumbent annual herb, 5-15(-25)cm; stems square, hairs patent, crinkled; leaves ovate, c. 5 × 3cm, acute, base cuneate-decurrent, marginal teeth shallow, c. 6; petiole c. 7cm; inflorescence 4-5cm, erect, internodes 0.5-1cm, 6-10 flowers per node, white bristle-hairy; flowers blue to purple-blue; corolla 6mm. Amongst rocks, submontane; 1636m.

Distr.: widespread in tropical Africa & Madagascar [tropical Africa & Madagascar].

IUCN: LC

Dom: Cheek sight record 24 9/2006.

Plectranthus glandulosus Hook.f.

A coarse, scrambling to erect, often robust, glandular and strongly aromatic herb, to ± 3.5m; leaves to 15cm long, glandular-punctate, margin with very uneven, rather small, double or treble crenations; inflorescence of copious loose panicles to ± 65cm long; mature calyx 9mm long; corolla violet; stamens declinate. Forest; 1600m.

Distr.: Mali to Bioko, Cameroon [upper & lower Guinea].

IUCN: LC

Dom: Dom photographic record 17 2/2006.

Plectranthus tenuicaulis (Hook.f.) J.K.Morton

Syn. *Plectranthus peulhorum* (A.Chev.) J.K.Morton

Slender, branched, annual herb, to ± 1m; stems pubescent; leaves shortly petiolate, lanceolate, acute, 0.5–6cm long, a third as broad, crenate, pubescent; inflorescence a panicle with lateral racemose branches; mature calyx 3–5mm; corolla 8–10mm, pale blue; stamens declinate. Forest, forest margins; 1636–1931m.

Distr.: tropical Africa [tropical Africa].

IUCN: LC

Dom: Cheek 13523 fl., 9/2006; 13586 fl., 9/2006; 13598 fl., 9/2006.

Local name: Nam-Meh (Cheek 13598).

Stachys aculeolata Hook.f. var. *aculeolata*

A perennial prickly herb; stems straggly, scrambling, rather slender with stiff prickly bristles and short hairs, sometimes with glandular hairs; leaves ovate-triangular, to 5.5 × 4.5cm, cordate; petiole as long as the lamina; inflorescence of distant, several-flowered verticillasters; calyx 7mm at maturity, campanulate; corolla c. 1cm long, white or pale pink; lower lip 3-lobed, central lobe larger, emarginate. Damp grassland, woods, swamps, forest margins; 1570m.

Distr.: Bioko, Nigeria, W Cameroon, Burundi, Rwanda, Congo (Kinshasa), Ethiopia, Uganda, Kenya, Tanzania [afromontane].

IUCN: LC

Dom: Dom photographic record 15 2/2006.

Vitex doniana Sweet

Tree, 10–20m; branches glabrous; leaves coriaceous, 5-foliolate; leaflets obovate to elliptic, middle ones 5–16 × 4–10cm; petiolule 1–2.5cm; inflorescence axillary or with terminal cymes, congested; peduncle 2–8cm long; fruits obovoid to subglobose, c. 1cm. Savanna; 1633–1640m.

Distr.: tropical Africa and the Comores islands [tropical Africa].

IUCN: LC

Dom: Cheek 13518 9/2006; Nana 143 4/2005.

Local name: Effileh (Cheek 13518). **Uses:** FOOD – Infructescences - fruit eaten (Cheek 13518).

Leeaceae

M. Cheek (K)

Fl. Cameroun 13 (1972).

Leea guineensis G.Don

Erect or sub-erect, soft-wooded shrub, to 7m; leaves bipinnate; leaflets opposite, imparipinnate, oblong-elliptic to 18cm long; flowers bright yellow, orange or red; fruits brilliant red, turning black. Forest and forest gaps; 1607m.
Distr.: Guinea (Bissau) & (Conakry), Sierra Leone, Ivory Coast, Togo, Ghana, Nigeria, Cameroon, Equatorial Guinea, Gabon, CAR, Congo (Kinshasa), Congo (Brazzaville), Burundi, Sudan, Uganda, Kenya, Tanzania, Malawi, Zambia, Angola, Madagascar, Reunion, Comoros and Mauritius [tropical Africa].
IUCN: LC
Dom: Cheek sight record 1 9/2006.
Local name: Bien (Cheek sight record 1).

Leguminosae-Caesalpinioideae

B. Mackinder (K)

Fl. Cameroun 9 (1970).

Caesalpinia decapetala (Roth) Alston

F.T.E.A. Leguminosae: Caesalpinioideae 36 (1967).
Spiny shrub or scrambler, to 6m; leaves bipinnate; leaflets 9-10 pairs, (obovate-)oblong, to 1.8×0.7cm; inflorescence an erect raceme; flowers pale yellow, 1.5cm diam., pod to 8.5cm. Secondary forest; 1580–1659m.
Distr.: Asia, widely cultivated in Africa [tropical Asia].
IUCN: LC
Dom: Cheek 13529 fl., 9/2006; Dom photographic record 32 2/2006.

Chamaecrista mimosoides (L.) Greene

Lock M., Legumes of Africa: 32 (1989).
Syn. *Cassia mimosoides* L.
Prostrate or more commonly erect herb or subshrub, to 1.5m; leaves paripinnate; leaflets 20-70 pairs, linear to linear oblong, $3-8 \times 1-1.5$mm; flowers yellow, 4-13mm; pod linear to linear oblong, up to 8cm. Grassland; 1831m.
Distr.: widespread in the palaeotropics [palaeotropics].
IUCN: LC
Dom: Cheek sight record 35 9/2006.

Senna septemtrionalis (Viv.) Irwin & Barneby

Lock M., Legumes of Africa: 39 (1989).
Syn. *Cassia laevigata* Willd.
Shrub, or small tree, to 3m; leaves paripinnae; leaflets 3-4 pairs, lanceolate to ovate, $4-11 \times 2-4$cm; flowers yellow, 1-1.5cm; pod subterete, up to 10cm. Farmbush & villages; 1630m.
Distr.: pantropical [pantropical].
IUCN: LC
Dom: Nana 139 4/2005.

Leguminosae-Mimosoideae

B. Mackinder (K), J.-M. Onana (YA) & M. Cheek (K)

Albizia glaberrima (Schum. & Thonn.) Benth.

Tree, c. 15m; leafy stems golden puberulent; leaves alternate, bipinnate, 30cm, pinnae in 6 pairs; rachis with raised gland 0.5-0.7mm diam. inserted on upper surface 1cm below the pinnae; pinnae leaflet pairs 8-14; leaflets falcate-oblong, c. 14 $\times$ 7mm, midrib diagonal. Forest edge, secondary forest; 1640m.
Distr.: widespread in tropical Africa [tropical Africa].
IUCN: LC

Dom: Pollard 1385 4/2005.

Albizia gummifera (J.F.Gmel.) C.A.Sm.

Syn. *Albizia gummifera* (J.F.Gmel.) C.A.Sm. var. *ealaensis* (De Wild.) Brenan
Tree, to 30m, crown flat; leaves bipinnate; leaflets numerous, usually glabrous, occasionally with hairs on the mid-rib and margins, not auriculate at base, up to 2×1.1cm; inflorescence capitate; calyx and corolla inconspicuous; stamens numerous, showy, up to 2.5cm long, greenish becoming red towards apex, fused into a tube, the free ends extending a further 5-7mm; pod compressed, coriaceous, glabrescent, becoming glossy, up to 18×3.2cm wide. Forest; 1570m.
Distr.: Nigeria to Congo (Kinshasa) [lower Guinea & Congolian].
IUCN: LC
Dom: Onana sight record 1 2/2006; Pollard sight record 3 4/2005.
Note: closely resembles *Albizia adianthifolia* (Schum.) W.F.Wight.

Entada abyssinica Steud. ex A.Rich.

Tree, to 10m; leaves bipinnate, 4-16 pinnae pairs, each pinna with 25-50 leaflet pairs; leaflets narrowly-oblong, less than 1cm long, pubescent, midvein not central; flowers subsessile, pale yellow, up to 2mm long, sweetly scented; pod compressed, the valves splitting transversely into 1-seeded papery segments leaving the persistant suture. Dry woodland and wooded grassland; 1490–1633m.
Distr.: widespread in Tropical Africa [tropical Africa].
IUCN: LC
Dom: Dom photographic record 31 2/2006; Cheek sight record 14 9/2006; Onana sight record 2 2/2006.

Newtonia camerunensis J.-F.Villiers

Bull. Jard. Bot. Natl. Belg., 60(1-2): 123 (1990).
Tree, 50cm diam. at base, fluted; leaves c. 25cm long; petiole 1.2–1.5cm; pinnae 8–10 pairs, gland between each pair; leaflets 25 per pinnae, 15×3.5mm, oblong, apex rounded, base rounded to slightly retuse, margin ciliate, upper surface glossy, midrib prominent; fruits $19–30 \times 1.8–2.3$cm, elliptic-oblong, straight or gently curved, apex rounded-apiculate; seeds winged, $4–8 \times 1.5–2.2$cm, elliptic to ovate. Forest; 1600–1640m.
Distr.: Cameroon (Bamenda Highlands and Bamboutos Mts) [W Cameroon Uplands].
IUCN: CR
Dom: Cheek 13433 9/2006; 13482 fr., 9/2006; Harvey 329 2/2006; Pollard 1394 4/2005.
Note: Cheek 13433 is possibly this taxon. The rachis glands compare well with the description/illustration of Villiers but the specimen is inadequate to determine to species (too immature) (Mackinder, x.2008). Cheek 13482 (pods only), geography and fruit size in accordance with field det. but material not sufficient to confirm determination (Mackinder, x.2008).

Leguminosae-Papilionoideae

B. Mackinder (K), J.-M. Onana (YA), B. Schrire (K) & M. Cheek (K)

Adenocarpus mannii (Hook.f.) Hook.f.

Shrub, to 5m; leaves 3-foliolate; leaflets very variable in shape, $5-8 \times 1.5-3.5$cm; flowers yellow, 9-14mm; pod oblong, up to 2.5cm, viscose-glandular indumentum. Grassland & forest-grassland transition; 1931m.
Distr.: Bioko, Cameroon, Congo (Kinshasa) & E Africa [afromontane].

IUCN: LC
Dom: Cheek 13575 9/2006.

Crotalaria incana L. subsp. *purpurascens* (Lam.) Milne-Redh.

F.T.E.A. Papilionoideae: 870 (1971); Polhill R., Crotalaria of Africa: 97 (1982); Opera Botanica 68: 168 (1983).
Herb, 0.6-1.2m; stems coarsely hairy; leaves obovate, 3-4cm long; flowers numerous, yellow with purple veins; fruit coarsely hairy. Forest edge; 1797m.
Distr.: tropical Africa [tropical Africa].
IUCN: LC
Dom: Cheek 13660 fl., 9/2006.

Crotalaria retusa L.

Erect herb, to 1.5m; leaves 1-foliolate, leaflet oblanceolate to oblong-obovate, 3.5-10 × 1.8-3.8cm; flowers pale yellow, reddish-purple-veined, 1.2-1.5cm; pod oblong, up to 2cm, inflated. Grassland, roadsides, disturbed ground and cultivations; 1633m.
Distr.: pantropical (origin probably Asiatic) [tropical Asia].
IUCN: LC
Dom: Cheek 13507 fl., 9/2006.

Crotalaria sp.

Shrub, to 1m; flowers yellow, beautifully veined black, apex of keel black. Grassland above forest; 1633m.
Dom: Cheek 13498 fl., 9/2006.
Note: Cheek 13498 was identified in the field and has yet to be verified (Harvey, iii.2009).

Dalbergia lactea Vatke

Scrambling shrub, to 3m; leaves imparipinnate; leaflets 9-13, alternate, oblong-elliptic or obovate, 3-7 × 1.5-2.5cm; flowers white or blue, 7-10mm; pod oblong, compressed, up to 15cm. Montane forest; 1810m.
Distr.: SE Nigeria to Congo (Kinshasa) & E Africa [afromontane].
IUCN: LC
Dom: Cheek 13655 fl., 9/2006.

Desmodium repandum (Vahl) DC.

Erect herb, to 1.3m; leaves 3-foliolate; leaflets rhombic-elliptic, 4.2-9.5 × 2.8-7.5cm; flowers orange-red or red, 8-11mm; pod strongly indented along the upper margin, up to 2.5cm. Forest & forest-grassland transition.
Distr.: palaeotropical [montane].
IUCN: LC
Dom: Pollard sight record 6 4/2005.

Eriosema parviflorum E.Mey. subsp. *parviflorum*

Erect or straggling herb, from woody base, up to 3m; leaves 3-foliolate; leaflets ovate or ovate-elliptic, 2.2-7 × 1.6-3.5cm; flowers reddish-yellow or yellow, 6-9mm; pod elliptic, up to 1.8cm, long pilose. Montane grassland & wooded grassland; 1600m.
Distr.: Cameroon to Zimbabwe, & E Africa [tropical Africa].
IUCN: LC
Dom: Cheek 13427 fl., 9/2006.

Eriosema sp. *of Dom*

Herb to 20cm; peduncle long; flowers reddish. Bushy grassland; 1780m.
Dom: Cheek 13543 fl., 9/2006.

Indigofera arrecta Hochst ex A.Rich

Shrub, 2m; stems black, with minute, appressed coppery hairs; leaves dense on stems, alternate, 5-9cm long, pinnately compound, 6-9-jugate; leaflets elliptic, c. 18 × 8mm; petiole 1.5cm; stipules persistent, slender, c. 8mm; racemes axillary,

numerous, 0.5cm, densely-flowered; flowers red, 4mm long. Fallow; 1633m.
Distr.: Senegal, Gambia, Mali, Guinea-Bissau, Guinea (Conakry), Ghana, Nigeria, Cameroon, extends to Saudi Arabia, South Africa [tropical & subtropical Africa].
IUCN: LC
Dom: Cheek 13512 fl., 9/2006.

Indigofera atriceps Hook.f. subsp. *atriceps*

Syn. *Indigofera atriceps* Hook.f. subsp. *alboglandulosa* (Engl.) J.B.Gillett
Herb, to 80cm; leaves imparipinnate; leaflets 2-7 pairs plus a terminal leaflet, 8-12 × 3-5mm; inflorescence an axillary raceme; flowers deep red, 5-7mm; pod narrowly oblong, up to 12mm, covered with glandular-tipped hairs. Montane grassland; 1931m.
Distr.: tropical Africa [afromontane].
IUCN: LC
Dom: Cheek 13595 fl., 9/2006.

Indigofera mimosoides Baker var. *mimosoides*

Shrub or scrambling herb, to 2m; somewhat woody, sparingly glandular; leaves imparipinnate; leaflets elliptic, 6–14 × 3–8mm; flowers red, 4–7mm; pod linear, to 1.6cm. Upland grassland, stream banks, forest margins; 1600m.
Distr.: Cameroon to E & SE Africa & Angola [tropical Africa].
IUCN: LC
Dom: Cheek 13408 fl., 9/2006.
Local name: Ndanso (Cheek 13408).

Kotschya strigosa (Benth.) Dewit & Duvign.

Robust herb, to 2m; leaves paripinnate, 5-12 pairs; leaflets linear, slightly falcate, 3-10 × 1-2mm; flowers bright blue, small yellow blotch at base of standard, 6-9mm; pod resembling a caterpillar, up to 8mm. Montane grassland; 1633m.
Distr.: Cameroon to Mozambique & to Angola, Indian Ocean [tropical Africa].
IUCN: LC
Dom: Cheek 13510 9/2006.

Leptoderris fasciculata (Benth.) Dunn

Climber; imparipinnate, 3-4 pairs of leaflets plus a terminal leaflet; leaflets elliptic to obovate elliptic 5-15 × 4-8cm; basic inflorescence form a spike with flowers clustered (fasciculate) along the axis, compounded up into a panicle up to 45cm; flowers purple or white; fruit flat, oblong, papery, winged, single-seeded. Forest; 1831–1980m.
Distr.: Guinea (Conakry) to Uganda [Guineo-Congolian].
IUCN: LC
Dom: Cheek 13636 9/2006; Pollard 1396 4/2005.
Note: Cheek 13636 is tentatively placed in this taxon, material sterile (Mackinder, x.2008).

Microcharis longicalyx (J.B.Gillett) Schrire

Bothalia 22(2): 166 (1992).
Syn. *Indigofera longicalyx* J.B.Gillett
Small herb, to 30cm; stem laxly pubescent; leaves elliptic, 25 × 5mm, mucronate, laxly pubescent; stipules paired, linear, persistent; inflorescence an axillary or terminal raceme to 5.5cm long; single bract at base of pedicel; calyx pubescent, deeply lobed; flowers red, blackish interior, 3mm long; pod linear, 20 × 2mm, pubescent, with raised margin. Grassland on rocky hills; 1931m.
Distr.: Guinea (Conakry), Sierra Leone, Nigeria, Cameroon, CAR [upper & lower Guinea].
IUCN: LC
Dom: Cheek sight record 31 9/2006.

Millettia drastica Welw.

Tree, to 15m; trunk circumference 30cm; leafy branches smooth, grey-brown; lenticels elliptic, slightly raised, pale yellow; leaves alternate, pinnately compound, 35cm; leaflets 9, oblong-elliptic, 12 × 4cm, long-acuminate, obtuse, secondary nerves c. 15 pairs, midrib raised above, glabrous except for sparsely hairy margin and petiolule; petiolule 3-4mm; spikes terminal, 30cm, 10-12-flowered; flowers pink-purple, white near base, 2cm, buds densely appressed golden-brown hairy; calyx cup-like; pods to 10cm (immature). Forest edge and scrub; 1650m.
Distr.: SE Nigeria, Cameroon, Rio Muni, Gabon, Congo (Kinshasa), Angola and Sudan [lower Guinea & Congolian].
IUCN: LC
Dom: Corcoran 4 4/2005.
Note: widespread but rare in W Africa (Cheek, vi.2009).

Pseudarthria hookeri Wight & Arn.

Erect herb, or sub-shrub, to 2m; leaves 3-foliolate; leaflets ovate or narrowly rhomboid to rhomboid, 5.5-13.5 × 2.5-6.5cm; flowers purple, blue, deep pink or white, sometimes pale yellow, 4-8mm; pod narrowly oblong, up to 2.5cm. Grassland, farmbush & grazed grassland; 1633m.
Distr.: widespread in tropical Africa [tropical Africa].
IUCN: LC
Dom: Cheek 13489 fl., fr., 9/2006; 13495 fl., 9/2006.

Sesbania macrantha Welw. ex Phill.

Lock M., Legumes of Africa: 462 (1989).
Shrub, 1.7-2m; leaflets imparipinnate, 10-24 jugate, leaflets oblong or oblong-elliptic, 9-24mm long; flowers deep yellow, 1.2-1.5cm; pod linear, slightly curved, up to 30cm. Wet places, swamps and stream margins; 1640m.
Distr.: widespread in tropical Africa [tropical Africa].
IUCN: LC
Dom: Onana 3637 2/2006.

Sesbania sesban (L.) Merr.

Sparsely branched soft-stemmed shrub, 4m, glabrous; leaves 26cm, pinnate, c. 30-jugate; leaflets oblong, c. 2.5 × 0.5cm, apex truncate, with mucro, base asymmetric, acute; petiolule 1-2mm; stipules lanceolate, 8mm; inflorescence axillary, 10-15cm; flowers 4-6, bright yellow, each 2.5cm across. Grassland above forest; 1633m.
Distr.: tropical Africa and Asia to N Australia, also St. Helena [palaeotropics].
IUCN: LC
Dom: Cheek 13508 9/2006.

Tephrosia vogelii Hook.f.

Erect shrub, 2-3m clothed with dense yellowish or rusty tomentum; leaflets of 5 or more pairs, oblanceolate, 4-7cm long; inflorescence a dense raceme; flowers conspicuous, red or reddish-purple. Forest, fallow; 1640m.
Distr.: tropical Africa [tropical Africa].
IUCN: LC
Dom: Onana 3636 2/2006.

Trifolium baccarinii Chiov.

Prostrate (occasionally ascending) herb, sometimes rooting at the nodes; leaves 3-foliolate; leaflets elliptic or obovate, finely toothed, 11-16 × 7-10mm; flowers purple or white, 3-4mm; pod broadly oblong, 2-3 × 1-1.5mm. Grazed grassland; 1797m.
Distr.: N Nigeria to Ethiopia & to Tanzania [afromontane].
IUCN: LC
Dom: Cheek 13564 fl., 9/2006.

Vigna gracilis (Guill. & Perr.) Hook.f. var. gracilis

Slender twining or semi-prostrate herb; leaves 3-foliolate; leaflets ovate, broadly elliptic or rhombic, 1-4.5 × 0.8-2.1cm; flowers pink or bluish, turning yellow; 9-16mm; pod linear, deflexed, up to 4cm. Wooded grassland, grassland & roadsides; 1636m.
Distr.: Widespread in W & WC Africa [Guineo-Congolian].
IUCN: LC
Dom: Cheek 13604 fl., 9/2006.

Lentibulariaceae

M.Cheek (K)

Utricularia andongensis Welw. ex Hiern

Annual terrestrial herb, to 10cm; leaves 4–5 in basal rosette, strap-shaped, 1–3 × 0.3cm; inflorescence often twining; bracts and bracteoles basifixed, ovate; calyx lobes subequal, 2–6mm; corolla c. 0.5cm, yellow; spur subulate, larger than lower lip. Rock face; 1797m.
Distr.: Guinea (Conakry) to Zambia [Guineo-Congolian].
IUCN: LC
Dom: Cheek 13554 fl., 9/2006.
Note: Cheek 13554 is tentatively placed in this taxon. It lacks fruits and seeds. Similar to *Utricularia spiralis* (Cheek, xi.2008).

Utricularia striatula Sm.

Epiphytic annual herb, 1–20cm, stoloniferous; leaves petiolate, blade transversely elliptic, to 1–5mm wide; inflorescence non-twining; bracts medifixed, oblong, to 1.5mm; upper calyx obcordate c. 2mm, lower one fifth as large; corolla 0.5–1cm, white; lower lobe flat, 5-lobed, upper inconspicuous; capsule with short ventral slit. Forest; 1931m.
Distr.: Guinea (Conakry) to Tanzania, India to New Guinea [palaeotropics].
IUCN: LC
Dom: Cheek 13609 fl., 9/2006.

Loranthaceae

R.M. Polhill (K)

Fl. Cameroun 23 (1982).

Loranthaceae indet. of Dom

Epiphytic shrub; stems densely lenticellate; flowers pink, faintly striped with darker lines. Grassland, side of path; 1931m.
Dom: Cheek 13599 fl., 9/2006.
Local name: Sah-Ton (Cheek 13599).

Agelanthus brunneus (Engl.) Balle & N.Hallé

Parasitic shrub, to 1m or more; hairs simple or absent; leaves ovate-lanceolate, 2.5–15 × 1.5–8cm, apex blunt, 3(–5)-nerved from base; flowers few–numerous, crowded at axils or older nodes, banded red, orange and white; corolla 3.2–4(–4.5)cm, 5-petaled, white or yellow over vents, conspicuously swollen basally from inception; petals erect. Forest & farmbush; 1500m.
Distr.: Senegal to W Kenya, S to Angola [Guineo-Congolian (montane)].
IUCN: LC
Dom: Corcoran sight record 1 2/2006.

Globimetula oreophila (Oliv.) Tiegh.

Parasitic shrub; twigs compressed; leaves lanceolate, ovate, 8–13 × 2.5–6cm, with 6–12 pairs of well-spaced curved-

ascending nerves; umbels 1–4, in axils, 8-21-flowered; peduncle 0.5–3.5cm; corolla red or red-purple, darkening apically as bud ripens, or red with a green or cream top, 2.5–3.5mm; basal swelling 5-shouldered. Forest-grassland transition; 1735m.
Distr.: SE Nigeria & Cameroon [W Cameroon Uplands].
IUCN: NT
Dom: Harvey 316 2/2006.
Note: possibly threatened by forest loss in the Bamenda Highlands.

Helixanthera mannii (Oliv.) Danser
Parasitic shrub, ± 1m; leaves elliptic-oblong to elliptic, ovate or oblong-lanceolate, 6–10(–14) × 1–4cm with 6–10(–12) pairs of fine lateral nerves; racemes terminal and in many axils, 2–10(–15)cm, 20–50-flowered; flowers white to pinkish; petals free, caducous, linear, (4–)8–18(–24)mm. Forest, often along rivers; 1490m.
Distr.: SE Nigeria to Uganda [lower Guinea & Congolian].
IUCN: LC
Dom: Harvey 324 2/2006.

Malvaceae

M. Cheek (K)

Hibiscus noldeae Baker f.
Subshrub, to 2m, spiny, very sparingly simple hairy; leaves orbicular in outline, c. 6cm diam., palmately 5-lobed almost to base, lobes elliptic-oblong, serrate, stipules caducous; flowers axillary, subsessile; epicalyx 8-10; bracts linear, 14mm, apex birfurcate; corolla 4cm, yellow. Disturbed forest; 1659m.
Distr.: Nigeria, Cameroon, Uganda, Congo (Kinshasa), Tanzania, Angola [afromontane].
IUCN: LC
Dom: Cheek 13530 fl., 9/2006.

Kosteletzkya adoensis (Hochst. ex A.Rich.) Mast.
Shrub, 0.3–0.9m or straggling in trees to 6m; subscabrid; leaves ovate or shallowly 3-lobed, to 5.5 × 3cm, acute, cordate, crenate; petiole 1cm; pedicel 1cm; epicalyx of c. 8 filiform bracts, 4mm; calyx 4mm; corolla pink, centre purple, 1.5cm; fruit strongly 5-ridged-winged. Forest edge; 1797m.
Distr.: tropical Africa [afromontane].
IUCN: LC
Dom: Cheek 13566 fl., 9/2006.

Pavonia urens Cav. var. *urens*
Subshrub; to 2m; densely persistent-pubescent on stems and leaves; leaves circular in outline, c. 15cm, more or less 5-lobed; fascicles axillary, 5–10-flowered; corolla pink, 1cm; mericarps 5; awns long exserted with retrorse spines. Forest edge; 1607m.
Distr.: Guinea (Conakry) to Madagascar [tropical Africa & Madagascar].
IUCN: LC
Dom: Cheek sight record 8 9/2006.

Urena lobata L.
Subshrub, 0.6m, stellate hairy; leaves elliptic, c. 6cm, slightly 3-lobed or entire, base rounded, teeth glandular, densely grey hairy below; petiole 2cm; flowers subsessile; epicalyx 5-lobed in upper half, 7mm; corolla pink, centre purple, 1cm; mericarps 5, spines with grapnel ends. Farmbush; 1831m.
Distr.: pantropical [pantropical].
IUCN: LC
Dom: Cheek sight record 34 9/2006.

Melastomataceae

M. Cheek (K) & J.-M. Onana (YA)
Fl. Cameroun 24 (1983).

Antherotoma naudinii Hook.f.
F.T.E.A. Melastomataceae: 1 (1975); Flora Zambesiaca 4: 220 (1978).
Erect, annual herb, c. 15cm tall; stems hirsute; leaves ovate to oblong-lanceolate; blade c. 2 × 1cm; inflorescence capitulate; flowers 4-merous; petals pink; anthers truncate; fruit a capsule. Short grassland, usually in damp places; 1797m.
Distr.: Guinea (Conakry) to Cameroon, Ethiopia to Angola & S Africa, Madagascar [afromontane].
IUCN: LC
Dom: Cheek 13563 fl., 9/2006.

Dissotis brazzae Cogn.
Erect herb, c. 1.5m tall; stems 4-winged; leaves c. 7.5 × 3cm, 9-nerved, ovate, apex acuminate; petiole c. 4mm long; inflorescence a terminal panicle; flowers 5-merous; petals pink to violet; fruit a capsule. Grassland, roadsides; 1600m.
Distr.: Guinea (Conakry) to Ethiopia, and to Zambia [tropical Africa].
IUCN: LC
Dom: Cheek 13426 fl., 9/2006.
Local name: Kehtonton (Cheek 13426).

Dissotis longisetosa Gilg & Ledermann ex Engl.
Ascending herb, 0.5m, setose; leaves opposite, ovate, c. 9 × 4.5cm, acute, truncate-rounded, upper surface with sigmoid hairs, lower part dilate, adnate to lamina; petiole 1cm; peduncle c. 20cm, robust, with 1-2 pairs reduced leaves, 1-2cm; capitula 2-3 flowered; hypanthia c. 2 × 1.5cm, emergences very short and densely hairy. Grassland; 1931m.
Distr.: Cameroon [uplands of Western Cameroon].
IUCN: VU
Dom: Cheek 13593 fl., 9/2006.

Dissotis perkinsiae Gilg
Fl. Cameroun 24: 30 (1983).
Shrub, 1-1.8m; leaves ovate to ovate-lanceolate, 5-8cm long; flowers 11cm diam.; petals 3cm long, reddish-violet or deep mauve-purple. Grassland; 1633–1640m.
Distr.: Togo to Uganda [Guineo-Congolian].
IUCN: LC
Dom: Cheek 13493 fl., 9/2006; Dom photographic record 8 2/2006.

Dissotis thollonii Cogn. ex Büttner var. *elliotii* (Gilg) Jacq.-Fél.
Fl. Cameroun 24: 28 (1983).
Syn. *Dissotis elliotii* Gilg var. *elliotii*
Erect herb, to c. 3m tall; stems, petioles and main inflorescence axis glabrescent, long hairs at nodes; leaves lanceolate, 17 × 3cm, 5-7 nerves; petiole c. 4cm long; inflorescence a panicle, buds enclosed by bracteoles; flowers 5-merous; petals purple, c. 1.5cm long; pedicel 1cm long; anthers c. 9mm long; fruit a capsule. Montane grassland; 1600m.
Distr.: Sierra Leone to Cameroon [upper & lower Guinea].
IUCN: LC
Dom: Cheek 13407 fl., 9/2006.
Local name: Kehtonton (Cheek 13407).

Meliaceae

M. Cheek (K)

Carapa grandiflora Sprague
Tree, 6–20m, glabrous; leaves to 1.2m, paripinnate, 4–7-jugate; petiole c. 15cm; leaflets oblong to oblong-obovate, c. 18 × 7cm, rounded, acute; petiolules c. 1cm; inflorescence a terminal panicle, c. 30cm; flowers white, c. 8mm; sepals and petals greenish; staminal tube white; disk orange; stigma white; fruit 5-valved, subglobose, c. 10cm, warty; seeds c. 3cm. Forest; 1831–1980m.
Distr.: Nigeria to Uganda [lower Guinea & Congolian (montane)].
IUCN: NT
Dom: Cheek 13653 9/2006; Pollard 1397 4/2005.
Note: *Carapa* is currently being revised, after which several additional taxa, further to the two accepted in FWTA, might be revealed. The Dom taxon is probably to become *C. oreophila* Kenfack when the revision is published (Onana, vi.2009).

Entandrophragma angolense (Welw.) C.DC.
Forest tree, to 55m; 3m girth; bole long and straight; crown open; buttresses blunt, broad, low; bark smooth, pale grey-brown to orange-brown with papery scales, scales flaking high up tree; slash dark red and pink; leaves paripinnate, clustered at ends of branches, 7–10-jugate; rachis 25–45cm; petioles 12–16cm, not winged, glabrous or puberulous; leaflets oblong-elliptic, 7–28 × 3–10.5cm, rounded and often mucronate, base rounded to obtuse, lateral nerves pubescent, 9–12 pairs; flowers yellowish; fruit 14–22 × 3.5–5cm, valves 2.5–3cm wide, 2.5–4mm thick; seeds winged. Forest and farmbush; 1600m.
Distr.: Guinea (Conakry) to Uganda & Sudan [Guineo-Congolian].
IUCN: VU
Dom: Pollard sight record 4 4/2005; Onana 3628 2/2006.
Note: the identification of Onana 3628 was made in the field and has yet to be verified (Harvey, iii.2009).

Melianthaceae

M. Cheek (K)

Bersama abyssinica Fresen.
Syn. *Bersama maxima* Baker
Syn. *Bersama acutidens* Welw. ex Hiern
Tree or shrub, 2–8m, glabrous; leaves alternate, c. 30cm, variable, imparipinnate; leaflets 5–6 pairs, densely pubescent or glabrous, glossy, oblong-elliptic, c. 15 × 5.5cm, apex acute, base obliquely obtuse, lateral nerves c. 10 pairs, sometimes serrate in upper half; petiolule 0.5cm, rachis more or less winged; petiole c. 12cm; stipules c. 1cm, intrapetiolar; inflorescence a terminal raceme to c. 40cm; rachis c. 7cm; pedicels 1cm; flowers white, 1cm; fruit magenta red, dehiscent, ovoid, 2cm; seeds 1cm, arillate. Forest; 1607m.
Distr.: tropical Africa [afromontane].
IUCN: LC
Dom: Cheek sight record 7 9/2006.

Menispermaceae

L. Pearce (K) & M. Cheek (K)

Stephania abyssinica (Quart.-Dill. & A.Rich.) Walp. var. ***abyssinica***
Slender glabrous liana, to 10m; leaves ovate to orbicular-ovate, 5–10 × 4–13cm, entire, dark green above, glaucous beneath; petiole 4–12cm; inflorescences 4–7cm diam., to 40cm long; pseudo-umbel on a single peduncle to 10cm; rachis fleshy, red; flowers green or purple. Forest; 1640m.
Distr.: tropical Africa [afromontane].
IUCN: LC
Dom: Pollard 1384 4/2005.

Stephania abyssinica (Quart.-Dill. & A.Rich.) Walp. var. ***tomentella*** (Oliv.) Diels
Slender liana, to 10m, densely pubescent, with fine brown indumentum; leaves ovate to orbicular-ovate, 5–10 × 4–13cm, entire, dark green above, glaucous beneath; petiole 4–12cm; inflorescences 4–7cm diam., to 40cm long; pseudo-umbel on a single peduncle to 10cm; rachis fleshy, red; flowers green or purple. Forest; 1570m.
Distr.: widespread in central and E Africa [afromontane].
IUCN: LC
Dom: Dom photographic record 33 2/2006.

Monimiaceae

B. Tchiengue (YA) & M. Cheek (K)

Fl. Cameroun 18 (1974).

Xymalos monospora (Harv.) Baill. ex Warb.
Shrub or small tree, 3–8(–25)m; leaves opposite, leathery, elliptic, c. 10 × 4cm, acute, serrate; inflorescences c. 4cm, below leaves; fruit elliptic, 1cm with apical knob. Forest & forest-grassland transition; 1800–1831m.
Distr.: SE Nigeria to E & S Africa [afromontane].
IUCN: LC
Dom: Cheek 13644 9/2006; Nana 153 4/2005.

Moraceae

M. Cheek (K), O. Sene (YA) & E. Ndive (SCA)

Fl. Cameroun 28 (1985).

Ficus lutea Vahl
Kew Bull. 36: 597 (1981); Fl. Gabon 26: 193 (1984); Fl. Cameroun 28: 206 (1985); Keay R.W.J., Trees of Nigeria: 297 (1989); Kirkia 13(2): 261 (1990).
Syn. *Ficus vogelii* (Miq.) Miq.
Epiphytic shrub, later a tree; leaves obovate or oblong-elliptic, 13–27cm, lateral nerves 5–6 pairs; petioles exfoliating, stipules caducous; figs sessile, axillary and below leaves, dense, globose, 1cm, densely brown hairy; bracts 2, 2mm. Farmbush; 1831m.
Distr.: Senegal to S Africa & Madagascar [tropical Africa and Madagascar].
IUCN: LC
Dom: Cheek 13635 9/2006.
Local name: Kengum (Cheek 13635).
Note: a common high-altitude species in the Bamenda Highlands.

Ficus oreodryadum Mildbr.
Fl. Cameroun 28: 200 (1985); Kirkia 13(2): 269 (1990).
Tree, to 30m, or epiphytic shrub, glabrous; leaves coriaceous drying dark brown below, oblanceolate, c. 20 × 5cm, obtuse, subacuminate, base acute, secondary nerves c. 13 pairs; petiole 5.5cm, exfoliating, stout; stipules caducous; figs axillary, sparse, subglobular, 1.3cm, slightly warty, glabrous, sessile, stoutly beaked; basal bracts 3, 2mm. Forest; 1633m.

Distr.: SW Cameroon to Burundi, Rwanda & Uganda [lower Guinea & Congolian (montane)].
IUCN: LC
Dom: Cheek 13465 fr., 9/2006.
Local name: Ndoh (Cheek 13465).

Ficus sur Forssk.
Syn. *Ficus capensis* Thunb.
Tree, 5–20(–30)m, but fruiting at only 5m; stem to 60cm dbh; leaves elliptic-oblong, c. 14 × 7cm, shortly acuminate, rounded to obtuse, margin with c. 5 well-marked serrations on each side, subglabrous; petiole c. 4cm; stipules caducous; figs on branches c. 15cm long on main branches or trunk apex, (lowest c. 6m from ground), c. 2cm diam.; peduncle 7mm; bracts 3, c. 1.5mm, whorled. Farmbush and secondary forest; 1640–1700m.
Distr.: Senegal to South Africa [tropical Africa].
IUCN: LC
Dom: Corcoran 11 4/2005; Pollard 1391 4/2005.

Ficus thonningii Blume
Syn. *Ficus dekdekena* (Miq.) A.Rich.
Syn. *Ficus iteophylla* Miq.
Epiphytic shrub, glabrous; leaves elliptic, elliptic-oblong or oblanceolate-elliptic, 7.5–13 × 3.5–5.5cm, shortly acuminate, obtuse, entire, lateral nerves c. 12 pairs, fine, quaternary nerves conspicuous; petiole 2.3cm; stipules caducous; figs axillary, amongst and below the leaves, dense, sessile, globose, c. 0.8cm; basal bracts 2, fused to form a bilobed, brown puberulent, saucer-shaped structure 0.7cm diam. Forest; 1640m.
Distr.: tropical and S Africa [tropical and S Africa].
IUCN: LC
Dom: Pollard 1392 4/2005.
Note: common in high altitude Bamenda Highlands.

Trilepisium madagascariense DC.
Bull. Jard. Bot. Nat. Belg. 47: 299 (1977); Fl. Cameroun 28: 103 (1985); Keay R.W.J., Trees of Nigeria: 302 (1989).
Syn. *Bosqueia angolensis* Ficalho
Tree, 20m, glabrous; leaves elliptic c. 10 × 4.5cm, acuminate, obtuse, entire, lateral nerves c. 5 pairs, basal pair acute; petiole 1cm; stipules caducous; inflorescence axillary at leafless nodes, ellipsoid, 1.5 × 0.8cm, glabrous; peduncle 1.5cm; flowers emerging from apical aperture in inflorescence, in cluster c. 0.5 × 0.5cm. Forest; 1600–1831m.
Distr.: Guinea (Conakry) to Congo (Kinshasa) [Guineo-Congolian].
IUCN: LC
Dom: Cheek 13413 9/2006; 13651 9/2006.

Myricaceae

M. Cheek (K)

Morella arborea (Hutch.) Cheek
Cheek, M. *et al.*, The Plants of Mount Oku and the Ijim Ridge, Cameroon: 149 (2000).
Syn. *Myrica arborea* Hutch.
Tree, 5–10m, densely shortly pubescent; leaves alternate, aromatic when crushed, oblong, c. 7 × 2.5cm, acute, broadly obtuse, spiny-serrate, c. 11 pairs lateral nerves; petiole 1cm; spikes axillary, 3cm; fruiting capitula 3mm. Forest edge; 1931m.
Distr.: Bioko & Cameroon [W Cameroon Uplands].
IUCN: VU
Dom: Cheek 13573 9/2006.

Myrsinaceae

T. Utteridge (K) & M. Cheek (K)

Ardisia dom Cheek
Cheek, M. *et al.* The Plants of Dom, Bamenda Highlands, Cameroon: 73 (2009).
Shrub, 1.5-2m; stems 4-angled, red scurfy; leaves elliptic or narrowly elliptic, rarely oblanceolate, 10-15 × 2.5-7cm, weakly acuminate, base acute, lateral nerves c. 10 pairs, margin serrate, leaf glands dense, 0.2-0.3mm diam., black, conspicuous, secondary to quaternary nerves raised, yellow; petiole 1-2cm; inflorescence axes to 2cm, curved; pedicel 8mm in fruit; fruit globose, 5mm, brown-green, shortly rostrate, densely covered in raised elliptic black glands. Forest; 1600–1831m.
Distr.: known only from Dom [W Cameroon Uplands endemic].
Dom: Cheek 13455 fr., 9/2006; 13624 fr., 9/2006.
Note: not matched at Kew with any other African species, but several described from Cameroon, e.g. *Ardisia polyadenia* Gilg from Kebo, have no surviving material. This may be a lost species rediscovered. The raised nervelets, large dense glands, and dense regular leaf teeth make this species distinct (Cheek, 12.xi.2008).

Ardisia staudtii Gilg
Bull. Jard. Bot. Nat. Belg. 49: 112 (1979).
Syn. *Afrardisia cymosa* (Bak.) Mez
Syn. *Afrardisia staudtii* (Gilg) Mez
Shrub, (0.5–)1.5–4(–5)m tall, glabrous; leaves elliptic to ovate, 90–180 × 30–70mm, glandular dots present on lower surface, very shallowly crenate; petioles 7–15mm; flowers in axillary fascicles; peduncles 2–5mm; 6–12 flowers per fascicle; pedicel 6–10mm; calyx c. 2.5mm wide, fimbriate margin; flowers white or pink, to 4mm with glandular spots/streaks; fruits globose, 3–6.5mm, red with red gland-dots. Lowland & submontane forest; 1590m.
Distr.: Nigeria to Congo (Kinshasa) & CAR [Guineo-Congolian (montane)].
IUCN: LC
Dom: Onana 3604 2/2006.

Maesa rufescens A.DC.
Prodr. 8: 81 (1884).
Shrub, or small tree, 3–10(–18)m; stem glabrous, dark brown with paler lenticels; leaves elliptic-ovate, 9–16 × 4–5cm, serrate, hairs on midrib (at base) of undersurface and occasionally on secondary veins; petioles 1–4cm, glabrous; inflorescence many branched, 3–9cm, profusely hairy (to 1mm); flowers white, to 2 × 2mm, subsessile; fruits globose, 3.5–4mm; pedicel to 2mm. Forest & secondary forest; 1600–1640m.
Distr.: Guinea (Conakry) to E & S Africa, Madagascar [afromontane].
IUCN: LC
Dom: Cheek 13411 fr., 9/2006; 13496 9/2006; Dom photographic record 9 2/2006.
Local name: Sem (Cheek 13411) & Seb (13496).

Myrtaceae

M. Cheek (K)

Eucalyptus spp. of Dom
Tree, 10-20m; bole pink-brown; leaves narrowly lanceolate, slightly falcate, c. 12 × 2.5cm, apex long-attenuate, base unequally obtuse, midrib pale below, lateral nerves numerous, inconspicuous; infloresences 5-7cm; partial-

peduncles flattened; flowers white, c. 5mm diam.; calyx c. 6 × 3mm, including slightly narrower stipe. Cultivated; 1636m.
Distr.: native to Australia [widely cultivated].
Dom: Cheek sight record 22 9/2006.
Note: many species of *Eucalyptus* have been introduced into Africa and elsewhere from E Malesia (few) and Australia. The description above is meant to be a general one to cover the various species that may occur in our area.

Eugenia gilgii Engl. & Brehmer
Shrub, or small tree, to 8m; leaves elliptic, 6-10cm long; inflorescence a very short raceme, central bracteate axis to 4mm long; flowers pink borne below leafy parts of shoots. Submontane forest; 1550–1735m.
Distr.: Cameroon [W Cameroon Uplands].
IUCN: CR
Dom: Cheek 13432 9/2006; Onana 3609 2/2006; Pollard 1381 4/2005.
Note: Cheek 13432 is tentatively determined. The leaves are too big (Cheek, xi.2008).

Syzygium guineense (Willd.) DC. var. *guineense*
Syn. *Syzygium guineense* (Willd.) DC. var. *littorale* Keay
Syn. *Syzygium guineense* (Willd.) DC. var. *macrocarpum* (Engl.) F.White
Tree, with fire resistant bark, 3-7m, glabrous; stems white, cylindrical, glabrous; leaves opposite, leathery, elliptic, c. 12 × 5cm, acumen 0.5cm, base acute, lateral nerves, c. 20 pairs, tertiary veinlets raised, black dotted; petioles 2cm; inflorescence terminal, 15 × 15cm, 10-20-flowered; flowers 8mm; petals and sepals caducous; stamens numerous, white; fruit elliptic, 1cm. Secondary forest and forest edges; 1580–1640m.
Distr.: widespread in tropical African savanna [tropical Africa].
IUCN: LC
Dom: Nana 147 4/2005; Onana 3611 2/2006.
Note: Nana 147 & Onana 3611 determined as *Syzygium guineense* var. *macrocarpum* (syn. of var. *guineense*) by Cheek, xi.2008.

Syzygium staudtii (Engl.) Mildbr.
Tree, 8–20m; bole white, usually with 3- numerous laterally-flattened root buttresses, arising up to 60cm above ground; stems near apex red when young, 4-ridged, glabrous; leaves (fruiting stems) elliptic 6–7 × 2–3.5cm, acute, secondary nerves numerous; petiole c. 1.2cm; juvenile leaves to 11 × 5cm, sometimes briefly acuminate; inflorescence terminal, 10cm, 10–30-flowered; flowers white, 0.6cm; fruit subumbellate, obovoid, 1cm. Forest; 1797m.
Distr.: Liberia to Cameroon [upper & lower Guinea (montane)].
IUCN: NT
Dom: Cheek 13661 9/2006.
Note: unjustifiably reduced to a subspecies of *S. guineense* by White. Likely to rate as VU when taxon better delimited.

Ochnaceae

M. Cheek (K)

Campylospermum flavum (Schum. & Thonn.) Farron
Bull. Jard. Bot. Brux. 35: 397 (1965).
Syn. *Ouratea flava* (Schum. & Thonn.) Hutch. & Dalziel
Shrub, 1.5-4m tall, evergreen, glabrous; leaves glossy, leathery, alternate, elliptic-oblong, c. 23 × 6cm, acute at base and apex, margin acutely toothed, 2 teeth per cm; lateral nerves c. 15 pairs, quaternary nervelets raised; petiole 2mm; inflorescence terminal, c. 15 × 10cm, with 2-3 lateral

branches; flowers clustered in 4-5s; pedicels persistent, 2mm; flowers 1.5cm wide; petals 5, yellow, stalked; sepals 5, red, free; fruiting calyx persistent, 5mm diam; fruitlets glossy black, torus 5mm. Forest; 1490–1700m.
Distr.: Guinea to Bioko, Cameroon and CAR [Guineo-Congolian].
IUCN: LC
Dom: Corcoran 10 4/2005; Harvey 323 2/2006; Nana 149 4/2005.

Olacaceae

M. Cheek (K)

Fl. Cameroun 15 (1973).

Strombosia scheffleri Engl.
Tree, to 33m; branchlets strongly angled; leaves ovate-elliptic or oblong 6–20 × 3–13cm, 5–8 main lateral nerves, venation distinct; petioles 1–3cm long; flowers greenish-yellow or white; petals 3–5mm; fruits obconical, c. 2cm. Forest; 1640m.
Distr.: SE Nigeria to Uganda & E Africa [tropical Africa].
IUCN: LC
Dom: Nana 145 4/2005.
Note: Nana 145 is probably this taxon. However, it might as easily be *Strombosia sp. 1 of Bali Ngemba*. Fruit needed to be certain but subcordate leaf suggests possibility. Both possible at this range (Cheek, x.2008).

Strombosia sp. 1 of Bali Ngemba
Harvey, Y. *et al.*, The Plants of Bali Ngemba Forest Reserve, Cameroon: 115 (2004); Cheek, M. *et al.*, The Plants of Kupe, Mwanenguba and the Bakossi Mountains: 355 (2004).
Tree, 6–20m tall; stems and leaves resembling *S. scheffleri*, but leaves usually cordate; pedicels 2mm; flower buds black, 4mm; petals white, 5mm, inner surface sparsely and very shortly puberulent; fruit 1cm, resembling Diogoa: turbinate, with a calyx-derived wing around the equator. Forest; 1600m.
Distr.: Nigeria (Chappal Waddi) & Cameroon (Kupe-Bakossi, Bali-Ngemba F.R., Bamboutos Mts, Dom) [W Cameroon Uplands].
Dom: Cheek 13464 9/2006.
Note: Cheek 13464 is probably this taxon (*Strombosia* sp. 1 of Bali Ngemba). Fruit needed to be certain but subcordate leaf suggests possibility. Could as easily be *Strombosia scheffleri*. Both possible at this altitude and range (Cheek, x.2008).

Oleaceae

M. Cheek (K)

Chionanthus africanus (Knobl.) Stearn
Bot. J. Linn. Soc. 80: 197 (1980); Keay R.W.J., Trees of Nigeria: 403 (1989); Hawthorne W., F.G.F.T. Ghana: 45 (1990).
Syn. *Linociera africana* (Knobl.) Knobl.
Tree, 6–12m; bark pale brown, lenticels concolorous; leaves oblong-oblanceolate, 19.5–22.7 × 7–9.7cm, acumen abrupt, 0.6–1.4cm, base acute, glabrous, midrib raised and bronze below, lateral nerves 11–12 pairs, domatia minute, tufted; petiole 1.2–2.2cm, swollen, flaking; panicles axillary, c. 5 × 1.5cm, appressed pubescent; calyx lobes ovate, 1mm; corolla lobes subulate, 0.5cm, free almost to base, white; stamens 2; filaments short; fruit ellipsoid, 1.9 × 1.4cm, grey-brown. Forest & forest edge; 1590m.
Distr.: Sierra Leone to Tanzania [Guineo-Congolian].
IUCN: LC

Dom: <u>Onana 3605</u> 2/2006.
Note: Onana 3605 has atypical leaves that are much smaller than the norm (Cheek, xi.2008).

Jasminum dichotomum Vahl
Liana, to 8 m; stems glabrescent to minutely puberulent; leaves elliptic, 7–9.5 × 3.5–4.7cm, apex acuminate-apiculate, base acute, glabrous; petiole c. 0.6cm, puberulent; inflorescence densely corymbose, terminal on lateral shoots, flowers numerous, fragrant; calyx 3mm long, lobes triangular; corolla tube 1.7cm, red; corolla lobes c. 8, oblanceolate, 0.7cm long, spreading, white. Forest edge; 1735–1831m.
Distr.: Senegal to Mozambique [afromontane].
IUCN: LC
Dom: <u>Cheek 13613</u> fr., 9/2006; <u>Harvey 310</u> 2/2006.

Jasminum pauciflorum Benth.
Liana, to 6m; rather sparsely branched, subglabrous; leaves with tufts of hairs in the axils of the main nerves beneath, 3-9 × 1.5-5cm, ovate; petiole 2-10mm; few-flowered cymes; peduncles to 20mm; pedicels 20-30mm long; calyx pilose, lobes to 7mm; corolla white; corolla tube to 27mm long; corolla lobes 6-8, to 20mm long. Forest; 1633m.
Distr.: tropical Africa [afromontane].
IUCN: LC
Dom: <u>Cheek 13515</u> fr., 9/2006.

Olea capensis L. subsp. *macrocarpa* (C.H.Wright) I.Verd.
Kew Bull. 57(1): 108 (2002).
Tree, to 20m, glabrous; bark with pale to concolorous lenticels; leaves coriaceous, elliptic, 9.5–11 × 3–3.5cm, apex acuminate, base attenuate, margin slightly revolute, lower surface minutely punctate, midrib raised below, yellow to reddish, lateral nerves 6–7 pairs, inconspicuous; petiole to 2cm; cymose-panicles terminal, c. 8 × 7cm, branching opposite, patent, flowers numerous; pedicels 1–3mm; calyx cupular, lobes triangular, <1mm; corolla lobes ovate, c. 3mm, spreading, white; stamens 2, spreading; anthers 1.5mm long, medifixed; fruit ellipsoid, c. 1.3 × 0.7cm, green, smooth. Forest; 1810–1831m.
Distr.: Guinea (Conakry) to Somalia, South Africa, Madagascar & Comoros [tropical Africa & Madagascar].
IUCN: LC
Dom: <u>Cheek 13619</u> 9/2006; <u>13657</u> fr., 9/2006.

Passifloraceae

E. Fenton (K), B. Nke (YA), M. Cheek (K) & L. Pearce (K)

Adenia lobata (Jacq.) Engl.
Climber, to 10m, older stems swollen, with lines of tubercles, glabrous; leaves drying black, ovate, c. 14 × 8cm, acuminate, cordate, palmately nerved; petiole gland paired; fruiting pedicel 1cm; fruit globose, angled 2 × 2cm; stipe 2cm. Forest; 1600–1831m.
Distr.: Senegal to Cameroon [upper & lower Guinea].
IUCN: LC
Dom: <u>Cheek 13442</u> 9/2006; <u>13632</u> fr., 9/2006.
Local name: Kon-Teh Cheek 13442).
Note: Cheek 13442, is tentatively placed in this taxon (Cheek, xi.2008).

Adenia rumicifolia Engl. & Harms
Meded. Land. Wag. 71(19): 154 (1971); F.T.E.A. Passifloraceae: 35 (1975).

Liana, to 20; older stems terete; leaves entire or shallowly lobed, 3-35 × 2-20cm; petiole 1.5-15cm; petiole gland paired; fruits 1-4 per inflorescence, pear-shaped, 3-8 × 1.5-4.5cm; seeds 40-150 per capsule, ellipsoid, to 6mm. Forest; 1831m.
Distr.: tropical Africa [tropical Africa].
IUCN: LC
Dom: <u>Cheek 13631</u> 9/2006.

Passiflora edulis Sims
F.T.E.A. Passifloraceae: 15 (1975).
Climber, to 10m; stems glabrous, tendrils simple; leaves deeply trilobed, c. 11 × 11.5cm, lobes elliptic, acute, base rounded, margins serrate, glabrous, shiny; petiole c. 2cm with paired glands near apex; fruit globular, 4cm diam., brown, fleshy. Villages; 1659m.
Distr.: native of S America, cultivated in Africa [neotropics].
IUCN: LC
Dom: <u>Cheek 13533</u> 9/2006.
Uses: FOOD – Infructescences - fruit eaten (Cheek 13533).

Piperaceae

Y.B. Harvey (K) & M. Cheek (K)

Peperomia fernandopoiana C.DC.
Epiphytic herb, c. 4m from ground; stems erect, branched, 30cm, drying black, glabrous; leaves alternate, ovate-lanceolate, c. 7 × 3.5cm, acumen long, acute; inflorescences terminal and axillary, 2–3 per peduncle, to 6cm. Forest, secondary forest beside water; 1831m.
Distr.: Sierre Leone to Kenya [tropical Africa].
IUCN: LC
Dom: <u>Cheek sight record 40</u> 9/2006.

Piper capense L.f.
Pithy shrub, c. 1(–5)m; peppery aroma emitted when crushed; leaves opposite, broadly ovate, c. 15 × 10cm, cordate, glabrous except on nerves; inflorescences of leaf opposed, single, erect, white spikes c. 3 × 0.5cm. Forest; 1831m.
Distr.: Guinea (Conakry) to South Africa [tropical & subtropical Africa].
IUCN: LC
Dom: <u>Cheek sight record 37</u> 9/2006; <u>Pollard sight record 5</u> 4/2005.
Note: see photograph of *Newtonia camerunensis*.

Piper guineense Schum. & Thonn.
Hemiepiphyte-climber, reaching 20m above ground; peppery when crushed; stem twining and rooting adventitiously; leaves ovate-elliptic, to 19 × 10cm, obliquely-obtuse at base; inflorescence single, leaf-opposed, 3cm. Forest; 1590m.
Distr.: Guinea (Bissau) to Uganda [Guineo-Congolian].
IUCN: LC
Dom: <u>Onana 3601</u> 2/2006.

Pittosporaceae

M. Cheek (K) & Y.B. Harvey (K)

Pittosporum viridiflorum Sims s.l.
Shrub, or small tree, to 12m; immature branches and petioles pubescent to glabrescent; leaves usually crowded towards the end of branches, 7-17 × 1.5-4cm, obovate, lanceolate or spathulate, acuminate, cuneate; petioles to 20mm; inflorescences paniculate, with pubescent branches; pedicels 3-10mm; sepals 1.2-3 × 0.8-1mm; petals to 7mm; capsule valves to 8mm diam.; seeds mostly 4 on each valve. Riverine forest, forest, bushland, swamp and humid woodland; 1831m.
Distr.: tropical Africa [tropical Africa].

IUCN: LC

Dom: Cheek 13611 fr., 9/2006.

Note: Cheek 13611 lacks flowers, so is impossible to determine further without flowers (Cheek, x.2008).

Polygalaceae

M. Cheek (K)

Polygala tenuicaulis Hook.f. subsp. ***tayloriana*** Paiva

Fontqueria 50: 207 (1998).

Erect annual herb, 30–80cm; stem wiry, pubescent, unbranched in basal half; leaves sessile, linear-lanceolate, c. 30 × 1.5mm, pubescent; main inflorescence a dense terminal raceme to 12cm, continuous, many-flowered, short subterminal spikes may also be present; pedicels 2mm; flowers pale pink to purple; lateral sepals obovate, to 6 × 4mm, puberulent towards the base; fruit 4mm long, puberulent; seeds with long hairs towards apex, shorter below. Grassland; 1931m.

Distr.: Nigeria & Cameroon [W Cameroon Uplands].

IUCN: NT

Dom: Cheek 13597 9/2006.

Polygonaceae

M. Cheek (K)

Polygonum nepalense Meisn.

Syn. *Persicaria nepalensis* (Meisn.) H.Gross

Bot. Jahrb. Syst. 49: 277 (1913).

Straggling annual herb; stems sparsely pubescent; ocrea 0.7cm, glabrous; leaves ovate-deltoid, 3.2–4.2 × 2.4–3.8cm, apex acute, base truncate, margin sub-crenulate, glabrous; petiole 0.5–1.5cm, winged; inflorescence capitate, 0.5 × 0.5cm with a subtending sessile leaf 1 × 0.6cm; peduncle to 4cm, sparsely glandular-pubescent towards apex; bracts ovate 3mm, apex acute; perianth white; nut lenticular, c. 2mm diam., brown. Farmbush & grassland; 1659m.

Distr.: tropical & S Africa, Madagascar, tropical Asia [paleaotropics & sub-tropics].

IUCN: LC

Dom: Cheek 13532 9/2006.

Uses: MEDICINES (Cheek 13532).

Note: there is a trend towards placing all the *Polygonum* species recorded in our area within the genus *Persicaria*; however we here follow Lebrun & Stork (2003) in maintaining *Polygonum*. The equivalent combinations for *Persicaria* are listed as synonyms.

Polygonum senegalense Meisn. forma. ***albotomentosum*** R.A.Graham

F.T.E.A. Polygonaceae: 19 (1958).

Syn. *Persicaria senegalensis* (Meisn.) Sojak forma. *albotomentosa* (R.A. Graham) K.L.Wilson

Kew Bull. 45(4): 630 (1990).

Syn. *Polygonum lanigerum* Meisn. var. *africanum*

Herb, to 1.2m, matted white-tomentose throughout; ocrea 1.7–3cm, apical cilia absent; leaves ovate-lanceolate, 15.5–27 × 3–6.5cm, base acute; petiole c. 1.5cm; spikes dense, to 7.5cm long, 1–several together; bracts c. 4mm, ciliate; perianth white-green; nut lenticular, 3.5–4mm diam., shining dark brown. Moist open areas; 1797m.

Distr.: tropical & S Africa, Madagascar & Egypt [].

IUCN: LC

Dom: Cheek 13667 fl., 9/2006.

Uses: ENVIRONMENTAL USES – Unspecified Environmenal Uses - a protection against thunder and lightning (Cheek 13667).

Primulaceae

M. Cheek (K)

Anagallis djalonis A.Chev.

Erect, annual herb, 3–10cm, stems 4-angled, glabrous; leaves alternate, ovate, 5 × 3mm, obtuse; petiole 2mm; flowers single, axillary; pedicel 5mm; petals 5, 2mm, actinomorphic; fruit circumscissile. Footpaths; 1659m.

Distr.: Guinea (Conakry) to Kenya [afromontane].

IUCN: LC

Dom: Cheek 13606 fr., 9/2006.

Proteaceae

M. Cheek (K)

Protea madiensis Oliv. subsp. ***madiensis***

Fire adapted shrub, to 2m; inflorescence a capitulum; bracts yellow, tinged pink at tips, densely tomentose; flowers numerous, whitish-yellow or pink; flower 'limb' glabrous. Upland savanna, grassland; 1633m.

Distr.: S Nigeria, W Cameroon, from Ethiopia S to Angola, Mozambique [afromontane].

IUCN: LC

Dom: Cheek sight record 12 9/2006.

Ranunculaceae

M. Cheek (K)

Clematis simensis Fresen.

Tall woody climber, to 20m, pubescent on young growth, becoming glabrous; leaves imparipinnate with 5-leaflets (reduced in association with the inflorescence); leaflets ovate to ovate-lanceolate, occasionally with 1-2 lobes, subglabrous below; inflorescence many flowered; pedicels 1-5cm; sepals 7-18mm, cream or white. Forest edge & farmbush; 1500m.

Distr.: Cameroon to South Africa [afromontane].

IUCN: LC

Dom: Corcoran sight record 4 2/2006.

Clematis villosa DC. subsp. ***oliveri*** (Hutch.) Brummitt

Kew Bull. 55: 104 (2000).

Erect, perennial herb, with stout rhizome; stems 0.7-1.5m, strongly striate, 1-5-flowered; leaves pinnate to bipinnate or trifoliate, lobes very irregular in shape; flowers solitary, 3.5-5cm diam., white, pink or mauve; achenes up to 10cm diam. Savanna; 1815m.

Distr.: Nigeria, W Cameroon, Congo (Kinshasa), Rwanda, Burundi, Sudan, Uganda, Kenya, Tanzania [afromontane].

IUCN: LC

Dom: Harvey 318 2/2006.

Rhamnaceae

M. Cheek (K) & J.-M. Onana (YA)

Fl. Cameroun 33 (1991).

Gouania longipetala Hemsl.

Climber, with tendrils; leaves ovate, to 7.5 × 4.5cm, subacuminate, base rounded to truncate, crenate-serrate,

lateral nerves 3–4 pairs, sparsely hairy only on nerves; petiole to 2cm; stipules 1mm; inflorescence axillary, spike-like 10–15cm; flowers in fascicles, white, 2mm; fruits 3-winged, 6mm. Forest; 1659m.
Distr.: Guinea (Conakry) to Congo (Kinshasa) [Guineo-Congolian].
IUCN: LC
Dom: Cheek 13534 9/2006.

Gouania longispicata Engl.
Climber, tendrillate; leaves alternate, ovate, c. 4.5 × 3cm, base truncate, lateral nerves 6-7 pairs, densely red hairy, serrulate; petiole 1.5cm; racemes axillary, erect, c. 10cm long; flowers white, 3mm diam. Grassland & forest edge; 1750m.
Distr.: S Nigeria to Mozambique [afromontane].
IUCN: LC
Dom: Corcoran 9 4/2005.

Rhamnus prinoides L'Hér.
Shrub, 2–3m, glabrous; leaves glossy, dark green, alternate, elliptic, 7 × 3cm, subacuminate, acute, finely serrate, lateral nerves 4 pairs; petiole 1cm; inflorescence of single axillary flowers; pedicel c. 8mm; flowers green, 2mm; berry globose, 7mm, purplish red. Forest edge; 1831m.
Distr.: Cameroon to South Africa [afromontane].
IUCN: LC
Dom: Cheek 13612 fr., 9/2006.

Rosaceae

M. Cheek (K) & Y.B. Harvey (K)

Alchemilla cryptantha Steud. ex A.Rich.
Herb; stems prostrate, stoloniferous throughout; leaves reniform, c. 1.5 × 2.5cm, 5-palmatifid to palmatilobed, median lobe with 7–11 teeth, dentate, sparsely hairy; flowers white, inconspicuous, c. 2mm across; 2–8 carpels. Grassland; 1711m.
Distr.: tropical and S Africa [afromontane].
IUCN: LC
Dom: Cheek 13539 9/2006.

Prunus africana (Hook.f.) Kalkman
F.T.E.A. Rosaceae: 46 (1960); Blumea 13: 33 (1965); Fl. Cameroun 20: 209 (1978); Fl. Ethiopia 3: 32 (1989); Keay R.W.J., Trees of Nigeria: 181 (1989).
Syn. *Pygeum africanum* Hook.f.
Tree, to c. 20m; leaves alternate, lanceolate, 3–6 × 6–15cm, serrate; petiole 2cm long, bearing 2 glands near apex, or at base of lamina; inflorescence a dense panicle; flowers white, 5mm diam.; fruit a drupe, succulent, red, c. 1cm diam. Forest-grassland transition; 1633m.
Distr.: tropical and subtropical Africa [Afromontane].
IUCN: LC
Dom: Cheek 13485 9/2006.
Local name: Ed Deh (Cheek 13485).

Rubus pinnatus Willd. var. *afrotropicus* (Engl.) Gust.
Scandent, prickly shrub, to 5m; leaves less than 2.5 times as long as broad, glabrous, not glandular below; inflorescences terminal or less often axillary, many-flowered, rachis densely-appressed with short silver velvety-hairs; petals inconspicuous or caducous; infructescence with many fewer than 100 drupelets. Forest-grassland transition; 1600–1931m.
Distr.: tropical and S Africa [afromontane].
IUCN: LC
Dom: Dom photographic record 24 2/2006; Cheek sight record 29 9/2006.

Rubiaceae

M. Cheek (K) & S. Dawson (K)

Chassalia laikomensis Cheek
Kew Bull. 55(4): 884 (2000).
Shrub, 2–3(–8)m; leaves narrowly elliptic, 4–12 × 1.5–4cm, acuminate, lateral nerves 7–10 pairs; stipules 4mm, with conspicuous yellow raphides; panicles terminal, 5 × 5cm, loosely branched; flowers white, 6–10mm long; fruits black, ovoid, 6–9mm long. Forest; 1600–1831m.
Distr.: S Nigeria & W Cameroon [W Cameroon Uplands].
IUCN: CR
Dom: Cheek 13454 9/2006; 13621 fr., 9/2006; Nana 152 4/2005; Pollard 1388 4/2005.

Coffea liberica Bull. ex Hiern
Evergreen tree, to 10m, glabrous; leaves broadly elliptic, 7-13 × 3.5-6(-10)cm, acuminate, base acute to cuneate, lateral nerves 8-12, domatia on nerve, pit-like; petiole 0.8-2cm; stipule 2-4.5mm, apiculate, midrib inconspicuous; flowers several per axil, subsessile; corolla tube 5-6mm; lobes 6, 15 × 6mm, white; fruit red, ellipsoid, 14-18 × 8-10mm. Forest; 1831m.
Distr.: Guinea (Bissau) to Uganda [Guineo-Congolian].
IUCN: LC
Dom: Cheek 13642 fr., 9/2006.
Note: the varietal status of Cheek 13642 is as yet unresolved (Cheek, xi.2008).

Cremaspora triflora (Thonn.) K.Schum. subsp. *triflora*
Shrub, or climber, 1–8m; stem shortly pubescent; leaves papery, elliptic, c. 8.5 × 4.5cm, acuminate, rounded, nerves 3–5; petiole 0.5–1cm; flowers white, subsessile, 1cm; fruits red, axillary, fasciculate, ellipsoid, 13 × 6mm; pedicel 1mm. Evergreen forest; 1600–1640m.
Distr.: tropical Africa [tropical Africa].
IUCN: LC
Dom: Cheek 13414 fr., 9/2006; Nana 148 4/2005.

Cuviera longiflora Hiern
Shrub, or tree, to 8m, glabrous; stems with ants; leaves papery, lanceolate-oblong, c. 27 × 10cm, acuminate, base subcordate or rounded, nerves 9–10; petiole 1cm; stipule sheathing, 5mm; flowers 10–20, in axillary panicles; peduncle 3–8cm; bracts and calyx lobes leafy; corolla green and white; tube c. 1cm; lobes c. 0.2cm; fruit ellipsoid, 9 × 3cm, brown, fleshy; pyrenes 5, c. 3 × 1cm. Evergreen forest; 1450m.
Distr.: Cameroon to Angola [lower Guinea].
IUCN: NT
Dom: Harvey 320 2/2006; Onana 3615 2/2006.

Diodia sarmentosa Sw.
F.T.E.A. Rubiaceae: 336 (1976).
Syn. *Diodia scandens* sensu Hepper
Straggling herb, to 5m, scabrid; leaves ovate or ovate-lanceolate, 2.5–5 × 1–2.5cm, pubescent below, scabrid above; petioles to 7mm; stipule cupular 1mm, with 5 aristae, 3mm; axillary fascicles sessile; flowers white, 4-merous; fruit ellipsoid, 3–4mm, didymous, dry. Forest; 1633–1797m.
Distr.: pantropical [pantropical].
IUCN: LC
Dom: Cheek 13519 9/2006; 13669 9/2006.
Uses: MEDICINES (Cheek 13669).

Keetia venosa (Oliv.) Bridson
Kew Bull. 41: 974 (1986); F.T.E.A. Rubiaceae: 914 (1991); Fl. Zamb. 5(2): 365 (1998).
Syn. *Canthium venosum* (Oliv.) Hiern
Woody climber, to at least 4m; leaves elliptic, c. 9 × 5cm with hairy domatia; petiole 1cm; stipules narrowly triangular to aristate, 1cm long; fruit orange, fleshy with 2 stones, 7mm across, in dense clusters. Forest; 1600m.
Distr.: tropical Africa [tropical Africa].
IUCN: LC
Dom: Onana 3621 2/2006.
Note: Onana 3621 is atypical. It lacks the reticulate quaternary venation and more (Cheek, xi.2008).

Mussaenda arcuata Lam. ex Poir.
Climber, glabrous; leaves elliptic, c. 10 × 5cm, acuminate, acute, nerves 4–6; petiole to 2cm; flowers numerous without bract-like calyx lobe; corolla yellow; tube 1.5cm; lobes 1cm, centre orange, maturing red; fruit ellipsoid, 1.5cm, seeds numerous. Evergreen forest; 1931m.
Distr.: Gambia to South Africa [tropical Africa].
IUCN: LC
Dom: Cheek 13602 fl., 9/2006.
Local name: Mlah locally (Cheek 13602). **Uses:** FOOD – Infructescences - fruits eaten (Cheek 13602).

Mussaenda erythrophylla Schum. & Thonn.
Climber, densely puberulent; leaves elliptic, to 1.5 × 7cm, acuminate, obtuse, lateral nerves 9-13 pairs; petiole to 2cm; flowers 10-20; calyx with one leaf-sized, bract-like, red lobe; corolla cream to orange; tube 2cm; lobes 0.5cm. Evergreen forest.
Distr.: Guinea (Conakry) to S Africa [tropical Africa].
IUCN: LC
Dom: Pollard sight record 2 4/2005.

Oldenlandia rosulata K.Schum.
F.T.E.A. Rubiaceae: 290 (1976).
Erect, slender, annual herb, to 35cm, glabrous; basal leaves rosulate, blades spathulate to elliptic, 3-7 × 1.3-3mm, obtuse; stem leaves filiform, 0.3-3cm × 0.3-2mm; stipule sheaths with 2 teeth each 0.5mm; inflorescence paniculate; peduncles to 2.5cm; pedicels to 3.5cm; calyx 4-toothed; corolla white to pink; tube 1.2-3.6mm; lobes 4; capsule subglobose, 1-1.8mm. Grassland; 1633m.
Distr.: Cameroon to Natal [tropical Africa].
IUCN: LC
Dom: Cheek 13486 fl., 9/2006.

Otomeria cameronica (Bremek.) Hepper
Herb; stems 0.3-1m long, usually prostrate; leaves ovate-elliptic or ovate-lanceolate, 2-8 × 0.4-4cm; flowers white; corolla-tube 3-5mm long; fruits ovoid. Forest and grassland; 1636m.
Distr.: Sierra Leone to Bioko & Cameroon [Guineo-Congolian (montane)].
IUCN: LC
Dom: Cheek 13520 fl., 9/2006.

Oxyanthus okuensis Cheek & Sonké
Cheek, M. *et al*, The Plants of Mount Oku and the Ijim Ridge, Cameroon: 159 (2000).
Shrub or small tree, 3-8m; branches horizontal leaf-blade elliptic or elliptic oblong, 9-13 × 2.5-6cm; petiole 5-8mm long; stipule oblong, apical third triangular, 12-19 × 3-9mm; inflorescences in only one axil, at alternate sides on successive nodes, erect, 20-50-flowered; flowers white; corolla tube 3-4cm long; lobes 7-8mm long; fruit with an apical rostrum; pedicel accrescent, 12-14mm long. Evergreen montane forest; 1590–1831m.

Distr.: Mount Oku & Dom [W Cameroon Uplands (endemic)].
IUCN: CR
Dom: Cheek 13424 fl., fr., 9/2006; 13620 fl., fr., 9/2006; Nana 154 4/2005; Onana 3602 2/2006.
Note: Cheek 13424 has a very small rostrum; Cheek 13620 lacks rostrate fruit (Cheek, x.2008).

Pauridiantha paucinervis (Hiern) Bremek.
Shrub, 2–4m; stems puberulent; leaves elliptic-oblong, c. 10 × 3cm, base acute, nerves 8, domatia usually absent; petiole 7mm; stipule subulate, 7 × 1mm; flowers 5–10, on 1–2 peduncles to 1.5cm, white; corolla tube 3mm; fruit 5mm, red or black. Evergreen forest; 1590m.
Distr.: Bioko & Cameroon [lower Guinea].
IUCN: NT
Dom: Onana 3606 2/2006.

Pavetta calothyrsa Bremek.
Ann. Missouri Bot. Gard. 83: 107 (1996).
Shrub, to 3m, glabrous; leaves opposite, glossy above, elliptic, 15-25 × 7-10cm, acumen indistinct, base acute-decurrent, lateral nerves 7 pairs, domatia absent; petioles 3-4(-6)cm; stipule sheath 4mm, awn 1.5mm; inflorescence terminal, c. 6 × 7cm, white puberulent, dense-flowered; flower bud 7mm long; fruit globose, 8mm, glossy; calyx lobes square, 4, erect, as long as tube. Edge of gallery forest; 1450m.
Distr.: Cameroon, Gabon & Congo (Kinshasa) [lower Guinea & Congolian].
IUCN: LC
Dom: Onana 3613 2/2006.
Note: we follow Manning (Ann. Miss. Bot. Gard. 83: 107 (1996) in restoring *P. calothyrsa* from synonymy under *P. nitidula*. This species has been reported from Gabon and Congo (Kinshasa) as well as Cameroon (Cheek, vi.2009).

Pavetta hookeriana Hiern var. *hookeriana*
Shrub, 2–3m, subglabrous; floriferous twigs 15cm; leaves papery, elliptic to 13 × 6cm, acumen 0–1cm, cuneate, lateral nerves 10 pairs, domatia arched, hairy, nodules not seen, tertiary venation inconspicuous; petiole 2cm; inflorescence to 10cm across; flowers to 100; calyx lobes rotund, 2mm; corolla white; tube 2–5mm; lobes 4–8mm. Forest-grassland transition; 1550m.
Distr.: Bioko & W Cameroon [W Cameroon Uplands].
IUCN: VU
Dom: Pollard 1378 4/2005.

Pavetta sp. A of Dom
Shrub, 2m tall, glabrous; leaves opposite, ovate or elliptic, 12-18 × 6-8.5cm, subacuminate, acute, c. 10 pairs lateral nerves, domatia transverse slits, puberulent; bacterial nodules as lines and blotches along midrib and secondary nerves; stipule sheath c. 2mm, corky; inflorescence minutely white puberulent, 14 × 7cm (including subtending stem); fruit grey, glossy, ellipsoid, 7mm; calyx not persistent. Forest and grassland; 1633m.
Dom: Cheek 13467 fr., 9/2006.
Note: Cheek 13467 has not been matched at K, but the material is scanty. More material, especially ripe fruits and flowers needed to identify (Cheek, xi.2008).

Pentas pubiflora S.Moore subsp. *pubiflora*
Herb, or subshrub, 1.5m; leaves narrowly elliptic, c. 7 × 2cm; petiole 1cm; flowers white; corolla tube c. 7mm long. Grassland, forest or forest edge; 1636m.
Distr.: Cameroon, Congo (Kinshasa), Uganda, Kenya [afromontane].
IUCN: LC
Dom: Cheek 13522 fl., 9/2006.

Psychotria moseskemei Cheek
Kew Bull. 57(2): 382 (2002).
Shrub, 2-4(-5)m, glabrous; stems dark purple, hollow, 3-5mm diam.; leaf-blades elliptic, rarely obovate, 6-14 × 3.5-6cm, acumen 2-4mm, base cuneate, secondary nerves and midrib pale red, nerves 8-14 on each side, domatia circular, glabrous; inflorescences terminal, 30-50-flowered, 2-6 × 2.5-4cm; corolla green or yellow-green; tube 1mm; lobes 5, 2 × 1.5-2mm; style bifid; fruit subglobose, 6-9mm diam. Forest; 1600m.
Distr.: NE Nigeria & Cameroon [W Cameroon Uplands (endemic)].
IUCN: CR
Dom: Cheek 13425 fr., 9/2006.

Psychotria peduncularis (Salisb.) Steyerm. var. suaveolens (Schweinf. ex Hiern) Verdc.
Kew Bull. 30: 257 (1975).
Syn. *Cephaelis peduncularis* Salisb. var. *suaveolens* (Schweinf. ex Hiern) Hepper
Shrub, 1.5cm, glabrous; leaves elliptic, c. 13 × 6cm, acumen 1cm, base acute, lateral nerves 12 pairs, impressed above, patent brown puberulent below, domatia absent; petiole 2cm; inflorescence terminal; peduncle 3cm, erect, puberulent in 2 longitudinal lines; capitula 2cm wide, enveloped in bracts; flowers white, densely packed, c. 2mm wide. Disturbed edge of forest; 1600m.
Distr.: Cameroon, Congo (Kinshasa), Ghana, Guinea, Uganda, Kenya, Tanzania, Malawi, Sudan, Zambia [tropical Africa].
IUCN: LC
Dom: Onana 3625 2/2006.

Psychotria psychotrioides (DC.) Roberty
Shrub, 2m, glabrous; leaves elliptic, to 14 × 7cm, subacuminate, cuneate, lateral nerves 12 pairs; petiole 2cm; stipule obovate, sheathing, 1cm; inflorescence sessile, capitate, 2cm diam.; flowers 10–15, sessile, white, each 8mm; fruit ellipsoid, 1.5cm; calyx foliose. Forest; 1650–1735m.
Distr.: tropical Africa [tropical Africa].
IUCN: LC
Dom: Corcoran 5 4/2005; Onana 3610 2/2006.

Psychotria schweinfurthii Hiern
Bull. Jard. Bot. Brux. 34: 146 (1964); F.T.E.A. Rubiaceae: 64 (1976).
Syn. *Psychotria obscura* Benth.
Erect shrub, 0.5-1.5m; stem glabrous or hairy; leaves opposite, elliptic, c. 11 × 5cm, acumen 1cm, lateral nerves 12-15 on each side, looping to form inframarginal nerves; petiole 0.5-1.5cm; stipules ovate, c. 1cm deeply bilobed, lobes acute; inflorescence pendant, appearing lateral, c. 7 × 7cm, with 3-4 pairs of patent branches subtended by narrowly elliptic bracts c. 8mm long; corolla 4mm long, white, 5-lobed. Forest; 1607m.
Distr.: Ivory Coast to Uganda [Guineo-Congolian].
IUCN: LC
Dom: Cheek sight record 3 9/2006.

Psychotria succulenta (Hiern) Petit
Shrub, 1–3m, drying dark brown, matt; leaves leathery, elliptic-oblong, 15 × 7cm, acumen 0.5cm, acute-obtuse, lateral nerves 12 pairs; petiole 1cm; stipule broadly elliptic, 1.5cm, entire; inflorescence loosely capitate, 3 × 3cm; peduncle 7cm; flowers white, 3mm; fruit ovoid, 5mm. Forest; 1570–1800m.
Distr.: Nigeria to Zimbabwe [afromontane].
IUCN: LC
Dom: Corcoran 7 4/2005; Onana 3632 2/2006.

Psychotria sp. A aff. calva Hiern
Harvey, Y. *et al.*, The Plants of Bali Ngemba Forest Reserve, Cameroon: 122 (2004); Cheek, M. *et al.*, The Plants of Kupe, Mwanenguba and the Bakossi Mountains: 383 (2004).
Shrub, 3m; stems chalky green; leaves papery, drying black above, grey below, bacterial nodules linear, along midrib; blade oblong-elliptic, 20 × 9cm, acumen 1.5cm, base obtuse-rounded, lateral nerves 10-12 pairs; petiole to 3cm; stipule ovate, aristate, bifid, to 8mm; inflorescence c. 4cm, 10–20-flowered; peduncle 2cm; infructescence 4–8cm; fruit subglobose, red, 6mm. Forest; 1600m.
Distr.: Kupe-Bakossi, Bali Ngemba FR, & Dom [W Cameroon Uplands].
Dom: Cheek 13439 fr., 9/2006.
Note: the determination of Cheek 13439 is uncertain since the K duplicate is poor (Cheek, xi.2008).

Psydrax acutiflora (Hiern) Bridson
F.T.E.A. Rubiaceae: 906 (1991).
Syn. *Canthium acutiflorum* Hiern
Syn. *Canthium henriquesianum* (K.Schum.) G.Taylor Meded. Land. Wag. 82(3): 271 (1982).
Climber, 5m; stems glabrous, 4-angled, lacking hooks or ants; leaves coriaceous, glossy, ovate, 9 × 4.5cm, acuminate, rounded, 4-nerved; petiole 0.5cm; stipule triangular, 4mm; flowers axillary, numerous, white, panicle contracted, 5mm; fruit compressed spherical, 8mm, aborted carpel evident. Evergreen forest; 1735m.
Distr.: SE Nigeria to Uganda [lower Guinea & Congolian].
IUCN: LC
Dom: Harvey 314 2/2006.
Note: is this taxon really distinct from *P. kraussioides*? Lateral nerves 4-5 pairs (Cheek, xi.2008).

Psydrax kraussioides (Hiern) Bridson
F.T.E.A. Rubiaceae: 907 (1991); Fl. Zamb. 5(2): 362 (1998).
Syn. *Canthium kraussioides* Hiern
Syn. *Canthium henriquezianum* sensu Hepper
Woody climber; leaves elliptic, shiny coriaceous, 9-13 × 4-6cm; stipules with a keeled lobe; flowers yellow-green; fruit 2-lobed (each almost spherical). Forest patches, roadsides and villages; 1831m.
Distr.: Guinea to Angola [Guineo-Congolian].
IUCN: LC
Dom: Cheek 13628 fr., 9/2006.
Note: Cheek 13628 has large fruits and leaves that have 5-6 main pairs of lateral nerves (Cheek, xi.2008).

Psydrax sp. 1 of Dom
Tree, 5m, glabrous, with ants in hollow stems; leaves elliptic, 12-13 × 6-8cm, acumen 0.5cm, base obtuse, lateral nerves 7-9 pairs, domatia absent; petiole 1.5cm; inflorescences 5 × 6cm, flat-topped; flowers white, foul-scented, c. 5mm long. Forest and grassland; 1633m.
Dom: Cheek 13484 fl., 9/2006.
Note: only known from Dom. Resembles *P. dunlapii*, which has 10+ nerve pairs, hairy domatia, larger leaves, also *P. subcordata* (pubescent midrib, subcordate leaves, circular, glabrous domatia). More specimens and fruit needed to resolve (Cheek, xi.2008).

Rothmannia hispida (K.Schum.) Fagerlind
Tree, 5–20m; stems, nerves and calyx hispid; leaves papery, elliptic, c. 15 × 6cm, acuminate, drying black; flowers white; calyx tube 1.5cm; limb 3cm; teeth 2cm; corolla basal tube 12 × 0.6cm; upper tube 3 × 2.5cm; lobes 1.5 × 1cm, outer surface grey silky hairy. Evergreen forest; 1600–1980m.
Distr.: Guinea (Conakry) to Congo (Kinshasa) [Guineo-Congolian].
IUCN: LC

Dom: Cheek 13453 fr., 9/2006; Pollard 1401 4/2005.

Rutidea sp. aff. decorticata Hiern
Climber, decorticating, reflexed spurs absent, young stems, stipules, petioles and midribs densely pubescent; stipules 1, filamentous, 1cm; blade oblong or oblanceolate, 9-12(-15) × (4-)5cm, acumen 0.5cm, obtuse to subcordate, lateral nerves c. 10 pairs; petiole 1.5-2cm; infructescence 10cm, terminal, paniculate; calyx 5 lobed; pedicel 3mm; fruit globose, 5mm, orange. Forest; 1600–1831m.
Distr.: known only from Dom [W Cameroon Uplands (endemic)].
Dom: Cheek 13412 9/2006; 13652 fr., 9/2006; Harvey 309 2/2006.
Note: a probable new species, so far only known from Dom. Flowering specimens are needed before it can be named (Cheek, xi.2008).

Rytigynia sp. A of Kupe
Cheek, M. *et al.*, The Plants of Kupe, Mwanenguba and the Bakossi Mountains: 389 (2004).
Syn. *Rytigynia neglecta sensu* FWTA 2: 186 (1963) non (Hiern) Robyns
Shrub, 2–3m, glabrous; leaves lanceolate, 7 × 3.5cm, acumen to 2cm, rounded, lateral nerves 4 pairs, domatia white hairy; petiole 3mm; stipule sheathing 6mm; inflorescences axillary, umbellate, 3–4-flowered, 3mm; pedicel 2–3mm; corolla white, tube 4–5 × 2.5mm; fruit globose, 1.2cm, 2-seeded; pedicel 8mm. Forest; 1831m.
Distr.: Cameroon [W Cameroon Uplands].
Dom: Cheek 13614 9/2006.

Sabicea tchapensis K.Krause
Bull. Jard. Bot. Nat. Belg. 50: 255 (1980).
Syn. *Sabicea efulenensis* sensu Hepper
Climber; stems with reflexed brown hairs 1-2mm long, very dense when young; leaves elliptic 12 × 4.5cm, acute, sparsely white long hairy below, c. 10 pairs of lateral nerves conspicuous below; petiole 1cm, densely hairy; stipule ovate, 8 × 5mm, reflexed, glabrous inside; inflorescences capitate, 2-5cm across, axillary; peduncle 1cm, hairy; bracts ovate, 1.5 × 1cm, acuminate, scattered hairy; flowers white, tubular, densely white appressed hairy. Forest; 1450m.
Distr.: Nigeria, Togo and Cameroon [lower Guinea].
IUCN: NT
Dom: Onana 3614 2/2006.

Spermacoce exilis (L.O.Williams) C.D.Adams ex W.C.Burger & C.M.Taylor
Fieldiana, Bot. n.s., 33: 316 (1993).
Syn. *Borreria ocymoides* (Burm.f.) DC.
Syn. *Spermacoce ocymoides* Burm.f.
F.T.E.A. Rubiaceae: 361 (1976).
Erect, annual herb, 5cm; internodes c. 1cm, with 2 lines of hairs; leaves elliptic, 4-8mm, acute, 3-4 lateral nerves each side, glabrous; petiole 1mm; flowers white, c. 2mm across, in sessile terminal clusters of 6-8; fruits dry, dehiscent 1mm diam. Grassland; 1711m.
Distr.: Senegal, Guinea (Conakry), Sierra Leone, Liberia, Ivory Coast, Ghana, Nigeria, Cameroon, CAR, Gabon, Equatorial Guinea, Uganda, São Tomé, Congo (Kinshasa), [Guineo-Congolian].
IUCN: LC
Dom: Cheek 13538 fl., 9/2006.

Spermacoce princeae (K.Schum.) Verdc. var. princeae
F.T.E.A. Rubiaceae: 362 (1976).
Syn. *Borreria princeae* K.Schum. var. *princeae*

Creeping sometimes scandent herb, to about 60cm; stems 4-angled, 3mm wide, glabrous; internodes 4-12cm; leaves ovate to ovate-lanceolate, 2-5 × 1-2cm, lateral nerves 5 on each side, deeply impressed above, prominent below, puberulent; petiole 3-5mm; stipule cup 6 × 6mm, apex truncate, with 10 equally spaced bristles 6-10mm long; flowers axillary, sessile; corolla white, 4-lobed, 8mm wide. Forest, forest-grassland transition; 1640m.
Distr.: Cameroon to Tanzania [Guineo-Congolian (montane)].
IUCN: LC
Dom: Nana 141 4/2005.

Spermacoce pusilla Wall.
F.T.E.A. Rubiaceae: 356 (1976).
Syn. *Borreria pusilla* (Wall.) DC.
Erect annual herb, 10–20cm, glabrous; stems wiry, terete; leaves linear 2.5 × 0.2cm, sessile; stipule cup 1mm, deep, apex truncate, bristles 6, each 2mm long; flowers axillary, sessile; corolla pink; lobes 4. Roadsides & villages; 1633m.
Distr.: tropical Africa & Asia [palaeotropics].
IUCN: LC
Dom: Cheek 13501 fl., 9/2006.

Spermacoce spermacocina (K.Schum.) Bridson & Puff
Kew Bull. 44: 138 (1989).
Syn. *Borreria saxicola* K.Schum.
Erect herb, 10-20cm; stem hairs patent, white, 1.5mm; internodes 3-6cm; leaves elliptic 3 × 0.8cm, sparsely long-hairy, acute, lateral nerves 4 on each side, conspicuous below, sessile; flowers terminal in sessile heads of 4-6, pink; corolla tube 8mm; lobes 4, 3mm long; style exserted, with pin-head. Forest-grassland transition; 1607m.
Distr.: Nigeria & Cameroon [lower Guinea (montane)].
IUCN: LC
Dom: Cheek 13402 fl., 9/2006.

Spermacoce sp. aff. spermacocina (K.Schum.) Bridson & Puff
Ascending herb, 10cm high; clump, 10cm wide; stems glabrous, red; leaves ovate-elliptic to 18 × 8mm acute, base constricted, subpetiolate, lateral nerves 3 on each side of the midrib, glabrous apart from margin; stipule cup c. 2mm deep, apex convex, 5 bristled; bristles 3mm, with patent scattered hispid hairs 1mm; flowers sessile in axils; corolla pink, 4-lobed; lobes 2.5mm; tube slender, 5mm; anthers long exserted, with style; stigma T-shaped. Rocky outcrop with wet and dry patches; 1797m.
Dom: Cheek 13551 fl., 9/2006.

Virectaria major (K.Schum.) Verdc. var. major
Weak-stemmed shrub, 60cm, terete, puberulent; leaves ovate-elliptic, to 12 × 6cm, acumen 1cm, base obtuse, abruptly decurrent, lateral nerves 10–12 pairs, sparsely softly hairy on both surfaces; petiole 2cm; stipule 5mm, entire; flowers in erect terminal clusters; calyx lobes linear, 1cm; corolla pale purple, 2 × 1.5cm; stamens 10, exserted 2cm. Forest, grassland edges; 1607m.
Distr.: Nigeria to Zimbabwe [afromontane].
IUCN: LC
Dom: Cheek sight record 2 9/2006.

Rutaceae

M. Cheek (K)

Fl. Cameroun 1 (1963).

Clausena anisata (Willd.) Hook.f. ex Benth.

Shrub or tree, 3–8m, non-spiny, puberulent, strongly aromatic; leaves imparipinnate, 15cm, 4–9-jugate; leaflets alternate, lanceolate-oblique, c. 6 × 2.5cm, acuminate, obtuse, lateral nerves c. 10 pairs; petiolules 1mm; panicle c. 12cm, slender; flowers white, 5mm; fruit indehiscent. Forest; 1600m.
Distr.: Guinea (Conakry) to Malawi [tropical Africa].
IUCN: LC
Dom: Cheek 13428 9/2006.
Local name: Doum Doum (Cheek 13428).

Zanthoxylum leprieurii Guill. & Perr.

Bull. Jard. Bot. Brux. 30: 403 (1960); Keay R.W.J., Trees of Nigeria: 324 (1989).
Syn. *Fagara leprieurii* (Guill. & Perr.) Engl.
Tree, 8m, spiny; leaves c. 30cm, c. 6-jugate, rachis sparingly spiny or smooth; leaflets drying dark brown, papery, oblong, c. 10 × 3cm, acumen 1.5cm, acute, cryptically serrate, lateral nerves 12–15 pairs, petiolule 1mm; panicle 9cm. Forest; 1831m.
Distr.: Senegal to Uganda [Guineo-Congolian].
IUCN: LC
Dom: Cheek 13650 9/2006.

Sapindaceae

M. Cheek (K)

Fl. Cameroun 16 (1973).

Sapindaceae sp. 1 of Dom

Tree, 3m, monopodial, lacking scent, hairy, hairs dense, patent, grey, 1.5mm; leaves to 70cm long, 5-jugate; rachis 38cm with cylindric opening 0.5mm wide (galls?); lateral leaflets subopposite, oblong-elliptic, basal leaflets with petiolule 2cm, blade 16cm, acumen 1cm, mucro, base acute, entire, lateral nerves 12 pairs, midrib hairy, pustulate below; distal leaflets to 28 × 8.5cm, lateral nerves 15 pairs; terminal leaflet obovate, c. 24 × 11cm; petiole 6cm; flowers and fruit unknown. Submontane forest; 1600m.
Dom: Cheek 13417 9/2006.
Note: this tree species, represented here by Cheek 13417, is unknown to us and may represent a new species, but further collections with flowers and fruits are needed to identify fully. The long hairs make it distinctive (Cheek, vi.2009).

Allophylus bullatus Radlk.

Tree, 15–18m; leaves trifoliolate; leaflets drying blackish-green above, brown below, secondary nerves 10–12 pairs, domatia conspicuous, white tufted, along midrib and secondary nerves, elliptic c. 19 × 8cm, long acuminate, cuneate, margin serrate; petiole c. 8cm; inflorescence in the leaf axils 10–21cm, branches 6–12 in the upper half to 10cm long; flowers white, 2mm. Forest; 1600m.
Distr.: Nigeria, Cameroon, Príncipe & São Tomé [W Cameroon Uplands (montane)].
IUCN: VU
Dom: Cheek 13443 9/2006.
Note: Cheek 13443 is an anomaly, domatia on the secondary nerves are present, but rarely (Cheek, xi.2008).

Allophylus ujori Cheek

Kew Bull. 64(3): 499 (2009).
Tree, or large shrub, to 10m; spines to 2cm; bark dull grey-brown; leaves trifoliolate; leaflets elliptic, to 13 × 6cm, base obtuse, apex acute or acumen to 1cm, inconspicuously serrate in distal half, to 10 secondary nerves per side; petiolule to 7mm; petioles to 10.5cm; panicles axillary, to 14cm with 4(-

6) branches each to 3cm long; fruit subglobose to 1 × 0.7cm (immature), single seeded. Submontane evergreen forest, especially near streams; 1550–1600m.
Distr.: SE Nigeria & NW Cameroon [W Cameroon Uplands].
IUCN: EN
Dom: Cheek 13434 9/2006; Pollard 1380 4/2005.
Note: this is the taxon erroneously published in "The Plants of Bali Ngemba Forest Reserve, Cameroon" as *A. conraui* Gilg ex Radlk. (Cheek 10432 & Ujor 30334).

Deinbollia sp. 2 of Kupe

Cheek, M. *et al.*, The Plants of Kupe, Mwanenguba and the Bakossi Mountains, Cameroon: 399 (2004).
Tree, 4m; leaves 60cm, 10-jugate; leaflets narrowly elliptic, c. 14–5cm, acumen 1cm, lateral nerves 16 pairs, brochidodromous; petiole 20cm, terete, white, upper edge black scurfy puberulent; panicle 18cm, terminal, densely red pubescent; flowers 4mm. Forest; 1600–1980m.
Distr.: Mt. Kupe & Dom, Cameroon [W Cameroon Uplands].
Dom: Cheek 13436 9/2006; 13625 9/2006; Pollard 1400 4/2005.
Note: Pollard 1400 is juvenile, so leaflets fewer, bigger, wider, stems more slender than mature flowering trees, 5-8m tall. See Cheek 8709 from Oku, also juvenile (Cheek, xi.2008).

Paullinia pinnata L.

Woody liana, to 25m; leaves c. 12cm, 2-jugate, rachis winged; leaflets elliptic, to 11 × 6cm, obscurely toothed, apex rounded; petiole c. 10cm; inflorescence tendriliform, as long as leaves, spicate; flowers white, 3mm; fruit red, 3-lobed, obovoid, 4 × 1cm, stipitate; seed white and red, 0.5cm. Forest & farmbush; 1450–1607m.
Distr.: tropical Africa & America [pantropical].
IUCN: LC
Dom: Cheek sight record 4 9/2006; Corcoran sight record 5 2/2006.

Sapotaceae

M. Cheek (K)

Fl. Cameroun 2 (1964).

Pouteria altissima (A.Chev.) Baehni

Pennington T., Gen. of Sapot.: 203 (1991).
Syn. *Aningeria altissima* (A.Chev.) Aubrév. & Pellegr.
Tree, to 50m with straight cylindrical bole and slightly buttressed base; bark pale greyish; young growth finely pubescent; leaf blades elliptic to oblong-elliptic, 5-16 × 3-17cm, obtuse, emarginate or acuminate, broadly cuneate; petioles to 1.5cm long; flowers clustered in axils of current leaves; pedicels 3-6mm long; sepals 3.5-5.5 × 2.5-4mm; corolla greenish-cream to pale yellow; tube to 3.5mm; lobes to 2mm; fruit red, obovoid to subglobose, to 2mm diam; seeds obovoid, to 1.5cm long. Lowland rainforest & riverine forest; 1831m.
Distr.: Guinea to Sudan, SW Ethiopia & E Africa [tropical Africa].
IUCN: LC
Dom: Cheek 13634 9/2006.
Local name: Dijor (Cheek 13634).

Scrophulariaceae

M. Cheek (K) & J.-M. Onana (YA)

Alectra sessiliflora (Vahl) Kuntze var. *senegalensis* (Benth.) Hepper
Slender, erect, roughly pilose herb, c. 30cm, usually drying black; leaves opposite, alternate within inflorscence, sessile, ovate, serrate, 10–50 × 10–20mm; flowers yellow, 5mm. Grassland; 1780m.
Distr.: Senegal to Mozambique [tropical Africa].
IUCN: LC
Dom: Cheek 13548 fl., 9/2006.

Lindernia abyssinica Engl.
Small perennial, erect or ascending, 5–10cm; leaves elliptic-oblong, 3–5 × 7–13mm, ciliate, rather fleshy and congested; calyx-lobes lanceolate; flowers blue; fruits c. 1cm, shortly beaked. Grassland; 1931m.
Distr.: W Cameroon, Sudan, Ethiopia, Uganda [afromontane].
IUCN: LC
Dom: Cheek 13587 fl., 9/2006.

Veronica abyssinica Fresen.
Prostrate, creeping herb, usually drying dark brown; stem branched from the base, pilose; leaves opposite, petiolate, ovate, serrate except towards base, 2–4 × 1–2cm; inflorescence a slender axillary peduncle; flowers blue or pinkish, paired or a few together, 8–10mm diam.; fruit bilobed, pubescent. Grassland, forest-grassland transition & roadsides; 1780m.
Distr.: Nigeria to Zimbabwe [afromontane].
IUCN: LC
Dom: Cheek 13546 9/2006.
Note: Cheek 13546 is the only record of this taxon at Dom. The identification was made in the field and has yet to be verified (Harvey, iii.2009).

Simaroubaceae

M. Cheek (K)

Brucea antidysenterica J.F.Mill.
Shrub, or tree, to 10m; leaves imparipinnate, 10-35cm long; leaflets 4-5, oblong-ovate to ovate-lanceolate, rusty tomentose beneath, 4.5-14 × 2-7cm, margins undulate; inflorescence an elongated panicle to 35cm long; flowers clustered, subsessile, green. Forest; 1600–1640m.
Distr.: tropical Africa [afromontane].
IUCN: LC
Dom: Onana 3623 2/2006; Pollard 1387 4/2005.

Solanaceae

M. Vorontsova (BM) & Y.B. Harvey (K)

Brugmansia × *candida* Pers.
Fl. Rwanda 3: 359 (1985).
Syn. *Datura candida* (Pers.) Safford
Shrub, or small tree, to 3m; leaves ovate-acuminate, to 24 × 12cm; flowers white, fragrant, pendulous funnel-shaped, 25-30cm long; fruit lemon-shaped. Farm hedge; 1500m.
Distr.: native of the Andes, cultivated throughout the tropics [pantropical].
IUCN: LC
Dom: Dom photographic record 11 2/2006.

Solanum aculeastrum Dunal
Tree, or shrub, to 7m, armed; white tomentum on all parts except upper surface of leaves; spines plentiful, straight or sharply recurved, to 15mm, compressed; leaves conspicuously deeply lobed, to c. 10 × 10cm; inflorescence lateral, axillary, densely brown-scurfy; flowers pinkish-white, c. 1cm across; fruits globose to c. 4 × 4cm. Farms, rocky grassland; 1500–1931m.
Distr.: widely distributed in the tropics, and often cultivated as a hedge plant [pantropical].
IUCN: LC
Dom: Cheek 13600 fr., 9/2006; Dom photographic record 12 2/2006.
Local name: Diam (Cheek 13600).

Solanum anguivi Lam.
Flora Zambesiaca 8(4): 93 (2005).
Syn. *Solanum distichum* Thonn.
F.T.A. 4(2): 223 (1906)
Syn. *Solanum indicum* L.
Coarse, tomentose, undershrub, to 2m, spiny or not; leaves elliptic, very shortly pubescent above, subtomentose beneath, to 10–16cm; inflorescence a raceme-like cyme; flowers white, c. 5mm; fruits erect, globose, red, 1–1.5cm diam. Forest edge; 1831m.
Distr.: W to E Africa, south to South Africa, and Arabian Peninsula [tropical Africa].
IUCN: LC
Dom: Cheek 13643 fl., 9/2006.

Solanum torvum Sw.
Shrub, to 3m; stems occasionally armed, densely stellate hairy; leaves large, ± elliptic, to 10–16 × 4–12cm, subscabrid, lobate-sinuate to subentire; inflorescence of corymbose cymes, 2–5(–14)-flowered; corolla white (rarely purple), to 2.5cm; fruit c. 1cm, globose, dirty brown, occasionally drying black. A common weed in farmbush or forest.
Distr.: pantropical [pantropical].
IUCN: LC
Dom: Pollard sight record 7 4/2005.

Sterculiaceae

M. Cheek (K)

Cola anomala K.Schum.
Tree, 15–20m; crown dense; stems bright white from waxy cuticle; leaves in whorls of 3(–4), simple, entire, elliptic, to 17 × 7.5cm, subacuminate, base rounded to obtuse, lateral nerves c. 7 pairs; petiole c. 2cm; stipules caducous; panicles 3cm in leaf axils; flowers yellow without red markings, 1.5cm; fruit follicles to 12cm, 2-seeded, with knobs and ridges. Forest; 1633m.
Distr.: Cameroon & Nigeria [W Cameroon Uplands].
IUCN: NT
Dom: Cheek 13468 fr., 9/2006.
Local name: Kola (Cheek 13468). **Uses:** FOOD – Seeds (Cheek 13468).
Note: leaves much larger than in flowering specimens since sapling, no doubt. Stems lack white wax - was ethanol treatment used? (Cheek, x.2008).

Dombeya ledermannii Engl.
Engl. Bot. Jahrb. xlv: 319 (1910).
Tree, 3.5-15m; leaf-blade suborbicular to ovate, slightly 5-lobed, 11-19 × 5.3-15cm; cordate; petiole 4-7cm long; inflorescence an axillary cyme; peduncle 3-7cm long; petals white, 0.8-1.3cm long. Forest edge; 1590–1640m.

Distr.: Bamenda Highlands [W Cameroon Uplands].
IUCN: CR
Dom: Dom photographic record 34 2/2006; Onana 3608 2/2006; Pollard 1389 4/2005.

Pterygota mildbraedii Engl.
Flora Zambesiaca 1(2): 517 (1961).
Tree, c. 50m; leaves alternate, ovate, c 15 × 15cm, base cordate; domatia conspicuous; petiole 10cm. Forest; 1633m.
Distr.: Bamenda Highlands and Albertine Rift [afromontane].
IUCN: NT
Dom: Cheek 13480 fr., 9/2006; Pollard 1402 4/2005.
Local name: Shem Mon Fuhm (Cheek 13480).
Note: wood not used (too soft), Cheek (pers. comm.).

Thymelaeaceae

J.-M. Onana (YA), M. Cheek (K) & Y.B. Harvey (K)

Fl. Cameroun 5 (1966).

Gnidia glauca (Fres.) Gilg
Fl. Gabon 11: 95 (1966); Fl. Cameroun 5: 69 (1966); Fl. Afr. Cent. Thymelaeaceae: 61 (1975).
Syn. *Lasiosiphon glaucus* Fresen.
Tree, to 15 m; trunk much-branched; leaves oblanceolate, very acute, glabrous, 5-8cm long; flowerheads numerous, subsessile, c. 5cm diam.; petals spathulate, surrounded by large ovate glabrescent bracts. Open woodland, forest-grassland transition; 1633–1815m.
Distr.: Cameroon, Ethiopia, Zambia and E Africa [afromontane].
IUCN: LC
Dom: Cheek 13506 9/2006; Dom photographic record 13 2/2006; Harvey 319 2/2006.

Tiliaceae

M. Cheek (K)

Triumfetta cordifolia A.Rich. var. *tomentosa* Sprague
F.T.E.A. Tiliaceae: 87 (2001).
Shrubby herb, to 3m, puberulent; leaves slightly 3-lobed, basal leaves drying black, ovate, c. 13 × 10cm, long-acuminate, base cordate to obtuse, coarsely dentate, lower surface completely obscured by dense but short, brownish-white hairs; inflorescence raceme-like, sparingly branched; flowers 8mm; fruit drying black, globose, 10mm including hooked bristles to 4mm, very sparsely to densely white hairy. Roadside; 1633m.
Distr.: Nigeria to Zimbabwe [lower Guinea, Congolian & S Africa].
IUCN: LC
Dom: Cheek sight record 20 9/2006;
Note: Cheek sight record 20 is tentatively placed in this taxon.

Ulmaceae

M. Cheek (K)

Fl. Cameroun 8 (1968).

Trema orientalis (L.) Blume
Syn. *Trema guineensis* (Schum. & Thonn.) Ficalho

Tree, to 8m; young stems densely pubescent; leaves variable, distichous, ovate-lanceolate, 6.5–13.5 × 2.8–5.3cm, apex acuminate, base truncate, margin serrulate, lateral nerves 4(–6) pairs, alternate above basal pair, upper surface scabrid, lower surface scabrid or sparsely pubescent to densely pubescent; cymes axillary, c. 10–20-flowered; peduncle 0–0.5cm; flowers white, c. 2mm; sepals broadly elliptic, obtuse, puberulent; fruit globose, 2–3mm diam., green, styles and sepals persistent. Forest & farmbush; 1931m.
Distr.: widespread in tropical Africa & Asia [palaeotropics].
IUCN: LC
Dom: Cheek sight record 27 9/2006.

Umbelliferae

M. Cheek (K)

Fl. Cameroun 10 (1970).

Agrocharis melanantha Hochst.
Fl. Cameroun 10: 52 (1970); F.T.E.A. Umbelliferae: 33 (1989).
Syn. *Caucalis melanantha* (Hochst.) Hiern
Sub-erect perennial herb, to 60cm; rootstock woody; robust stems rounded, finely ridged; leaves bipinnate on rachis 7–15cm, c. 6 pairs of pinnae, each divided into c. 3 pinnules, serrate to approx. half their width, blades finely pubescent; petioles sheathing, membranous at node, densely pubescent; umbel dense, globular, terminal, c. 1.5–2cm diam.; peduncle to 20cm, pubescent; involucral bracts numerous, lanceolate, ciliate, 3–4mm; c. 12 subsessile flowers per umbel, each with a subtending involucel of lanceolate bracts; corolla white; fruits green, ellipsoid, c. 5 × 3mm, ridged, with reflexed barbed bristles along ridges, ciliate hairs between ridges. Montane grassland; 1659m.
Distr.: Cameroon, Bioko, Congo (Kinshasa), E Africa [afromontane].
IUCN: LC
Dom: Cheek 13605 fr., 9/2006.
Note: recognised by some authors (e.g. Lee (2002): Israel J. Pl. Sc. 50(3): 211) as a synonym of *A. gracilis* Hook.f.

Centella asiatica (L.) Urb.
Creeping perennial herb; long glabrous internodal stolons to 10cm, nodal rooting; petioles 5–20cm, pubescent particularly when young, sheathing, with subtending leafy lanceolate stipule c. 5mm long; blade reniform with regular crenate margin, non-lobed, glabrous, c. 2–3cm diam.; umbel 3–5mm; peduncle 1–1.5cm, pubescent; 1–5 umbels per node, subtended by leafy bract; 3–4 subsessile flowers per umbel;, petals pink-purple, subtended by 2 pubescent bracts 1–2mm; fruits ellipsoid, truncate at apex, 2(–4) × 1(–3)mm, with reticulate sculpturing. Damp grassland; 1636m.
Distr.: pantropical [montane].
IUCN: LC
Dom: Cheek sight record 21 9/2006.

Cryptotaenia africana (Hook.f.) Drude
Upright rhizomatous herb, 0.3–1m tall; stems circular, ridged, glabrous, to 2mm diam.; leaves concentrated towards base, sheathing; petioles 6–10cm glabrous; compound, biternate or with ternate terminal pinna and lobed lateral pinnae; petiolules to 2.5cm; leaflets c. 3.5 × 2cm, ovate, irregularly dentate, pilose especially on adaxial surface; cauline leaflets sub-lanceolate, 3 × 0.5cm; umbels sparse, florets borne on fine peduncles/pedicels; primary peduncles to 2cm; pedicels to 1cm, involucre absent; corolla white; fruits 2 × 1mm, ellipsoid to ovoid, glabrous, green, with persistent reflexed styles. Forest; 1600m.
Distr.: Cameroon, E Africa [afromontane].

IUCN: LC
Dom: Cheek 13422 9/2006.

Sanicula elata Buch.-Ham.
Upright herb, 0.5(–1)m; short stolon at base; upright stems c. 2mm diam., glabrous, ridged; basal rosette of 2–4 leaves on petioles, to 15cm, 3(–5)-lobed almost to base, with irregular sublobing and dentate-mucronate margins, c. 6 × 8cm, glabrous; cauline leaves smaller on petioles, to 5cm; inflorescence cymose, bifurcating 3–4 times with a central cyme on peduncle, 1–1.5cm long; lateral primary peduncles 5–6cm, secondary peduncles c. 1–2cm; each floret of 2–3 sessile flowers; florets and peduncles subtended by lanceolate bracts, 6–12mm; flowers 1–2mm; corolla white; fruits ellipsoid, 3 × 2mm, covered in hooked bristles, green. Forest including stream edges; 1600m.
Distr.: tropical & South Africa, Madagascar, Comores & temperate Asia [montane].
IUCN: LC
Dom: Cheek 13459 fl., 9/2006.

Urticaceae

C.M. Wilmot-Dear (K), J.-M. Onana (YA) & M. Cheek (K)

Fl. Cameroun 8 (1968).

Boehmeria macrophylla Hornem.
F.T.E.A. Urticaceae: 44 (1989); Fl. Zamb. 9(6): 108 (1991).
Syn. *Boehmeria platyphylla* D.Don
Shrub, to 2(–3)m; branches glabrous, except when young; leaves opposite, anisophyllous, ovate, 10–13.5 × 5.5–9cm, acuminate, base acute to rounded, margin serrate, basal lateral nerves prominent, upper surface sparsely pubescent, cystoliths punctiform, lower surface glabrescent; petiole to 6cm; spikes axillary, 7–50cm, whip-like, with glomerules of flowers spaced 1–10mm apart; male glomerules 1–2mm; female 2–3mm. Forest & forest edge; 1750m.
Distr.: tropical Africa & Madagascar, tropical Asia to SW China [palaeotropics].
IUCN: LC
Dom: Corcoran 8 4/2005.

Droguetia iners (Forssk.) Schweinf.
Perennial herb or undershrub, to 1.8 m; stems trailing and ascending; leaves opposite, ovate, 3-nerved, with dot-like cystoliths above, sparsely pubescent, 1.8-2.5cm; inflorescences axillary, androgynous, bowl-shaped or campanulate involucres. Forest understorey; 1600m.
Distr.: Bioko and SW Cameroon to E and South Africa, Yemen Highlands [afromontane].
IUCN: LC
Dom: Cheek 13451 9/2006.
Local name: Kin Fieh (Cheek 13451).

Elatostema monticola Hook.f.
Herb, to 30cm; leaves 3(–6) × 1.5cm, apex acute to obtuse, base rounded distally, cuneate proximally, marginal teeth 9–10 distally, c. 6 proximally, blade sparsely pubescent; stipules 2–4mm; inflorescence sessile, 7mm wide; bracts oblong, 2.5mm, ciliate. Rocks & forest floor; 1600m.
Distr.: Bioko & Cameroon to E Africa & Zimbabwe [afromontane].
IUCN: LC
Dom: Cheek 13462 fl., 9/2006.

Elatostema paivaeanum Wedd.
Herb, to 50cm, rarely branched, stipules conspicuous, lanceolate, 7–10mm; leaves drying green-black, large, 7.5–16 × 3–5.5cm, highly aymmetric with distal base subcordate, proximal base cuneate, marginal teeth 12–18 distally, 9–14 proximally, cystoliths dense, conspicuous; inflorescence sessile, c. 13cm wide; bracts broadly ovate (male) to lanceolate (female), c. 5mm, ciliate; bracteoles pilose, clearly so in female inflorescence. Forest; 1600m.
Distr.: Guinea (Conakry) to E Africa & Malawi [tropical Africa].
IUCN: LC
Dom: Cheek 13447 fl., 9/2006.

Pilea rivularis Wedd.
Fl. Cameroun 8: 163 (1968); F.T.E.A. Urticaceae: 29 (1989); Fl. Zamb. 9(6): 98 (1991).
Syn. *Pilea ceratomera* Wedd.
Erect herb, to 60cm; stems with linear cystoliths; stipules prominent, oblong, 7mm; leaves to 7.5 × 5cm, base rounded, margin serrate-crenate, cystoliths linear; inflorescence a dense axillary cluster to 1.5cm diam. Forest including rocky stream margins; 1600m.
Distr.: tropical & subtropical Africa [afromontane].
IUCN: LC
Dom: Cheek 13450 fl., 9/2006.
Local name: Kimbin (Cheek 13450).

Pilea tetraphylla (Steud.) Blume
Herb, to 20cm; internodes decreasing and leaf size increasing towards apex; leaf pairs decussate; blade ovate to 3 × 2.2cm, apex scarcely acuminate, margins deeply serrate-apiculate, cystoliths linear, sparsely long-hairy, uppermost leaves subsessile; inflorescence appearing terminal, a corymb of 4 uneven parts, 2–3cm broad, dense. Forest, villages; 1931m.
Distr.: tropical Africa [afromontane].
IUCN: LC
Dom: Cheek 13610 fl., 9/2006; Cheek sight record 32 9/2006.

Verbenaceae

M. Cheek (K)

Lippia rugosa A.Chev.
Robust woody perennial, to 3m; stems distinctly appressed-pubescent; leaves ternate, to c. 10 × 3cm, oblong-lanceolate, upper surface scabrid, venation prominently rugose-reticulate; inflorescences spreading, much-branched corymbose cymes; flowers small, whitish. Grassland, savanna; 1633m.
Distr.: Guinea (Conakry), Nigeria, Cameroon [upper & lower Guinea].
IUCN: LC
Dom: Cheek sight record 18 9/2006.
Note: Cheek sight record 18 is a tentative field determination.

Vitaceae

L. Pearce (K)

Fl. Cameroun 13 (1972).

Cissus aralioides (Welw. ex Baker) Planch.
Fl. Cameroun 13: 88-91 (1972).
Climber, to 15m; stems terete, succulent, glabrous, drying yellow-green; tendrils simple or bifid; leaves palmately (3–)5-foliolate, 11.5–14(–22) × 4-8(–12)cm, central leaflet obovate, 9-18 × 4.8-5.5cm, apex acuminate, base acute, margin finely toothed, lateral leaflets asymmetric, ovate-elliptic, glabrous; petiole 7–14.5cm; inflorescence a compound, many-flowered cyme to 25 × 9.5cm; peduncle glabrescent; flowers 0.35cm; calyx cupular, puberulent;

corolla buds rounded at apex; pedicels 0.6cm, minutely puberulent; fruit ellipsoid, 2.5 × 1.5cm, glabrous, green-red; seeds flattened, ellipsoid, 1.8 × 0.9cm, smooth. Forest & farmbush; 1831m.

Distr.: Senegal, Guinea (Bissau), Guinea (Conakry), Sierra Leone, Liberia, Ivory Coast, Ghana, Nigeria, Cameroon, Equatorial Guinea, Gabon, CAR, São Tomé, Congo (Kinshasa), Sudan, Uganda, Kenya, Tanzania, Mozambique & Angola [tropical Africa].

IUCN: LC

Dom: <u>Cheek 13623</u> 9/2006.

Cyphostemma rubrosetosum (Gilg & Brandt) Desc.

Fl. Cameroun 13: 56 (1972).

Syn. *Cissus rubrosetosa* Gilg & Brandt

Herbaceous climber; stems cylindrical, densely glandular-hairy, hairs to 5mm, red, tendrils bifid; leaves papery, palmately 5-foliolate, 12.5 × 14cm; leaflets obovate, central 12 × 4.8cm, apex shortly acuminate, base cuneate, margin crenate-dentate, sparsely pilose along veins; petiole 7–10cm, pilose; compound cyme, c. 11 × 17cm; peduncles pubescent; flower 0.4cm; calyx cupular; corolla pinched at the centre, apex truncate, glabrous; pedicels 0.3cm, pubescent; fruit subglobose, 0.5cm diam., glabrous; seeds subglobose, c. 4mm, striate. Farmbush & open forest; 1607m.

Distr.: Guinea (Conakry) to CAR [upper & lower Guinea].

IUCN: LC

Dom: <u>Cheek 13399</u> fr., 9/2006.

Local name: Kunchien (Cheek 13399). **Uses:** MEDICINES (Cheek 13399).

MONOCOTYLEDONAE

Agavaceae

M. Cheek (K)

Agave sisalana Perrine
Fl. Zamb. 13(1): 33 (2008).
Monocarpic, rosette-forming, short-lived perennial; leaves sessile, erect, 120-200 × 10-15cm, linear, apex a spine, 1-1.5cm long; inflorescence terminal, paniculate, 5-6m tall with 25-40 flowering branches, each up to 30cm long with up to 40 flowers; fruits capsular; numerous bulbils often form on the inflorescence after flowering. Cultivated; 1633m.
Distr.: Mexico (widely cultivated elsewhere) [neotropics].
IUCN: LC
Dom: Cheek sight record 19 9/2006.

Amaryllidaceae

Y.B. Harvey (K) & M. Cheek (K)

Fl. Cameroun 30 (1987).

Scadoxus multiflorus (Martyn) Raf.
Fl. Cameroun 30: 8 (1987).
Syn. Haemanthus multiflorus Martyn
Bulbous herb, 25–80cm; bulb cylindrical, c. 2 × 1.5cm; leaves expanding after flowering, ovate-lanceolate, to 25 × 8cm, base attenuate; inflorescence lateral, 7–25cm, globose, many-flowered; flowers scarlet; pedicels 1.5–3.5cm. Forest edge & semi-deciduous forest; 1735m.
Distr.: Senegal to Somalia & to S Africa, also Yemen [tropical Africa].
IUCN: LC
Dom: Dom photographic record 25 2/2006.

Scadoxus pseudocaulus (Bjornst. & Friis) Friis & Nordal
Fl. Cameroun 30: 6 (1987).
Syn. Haemanthus sp. A sensu Hepper
Herb, to 80cm, rhizome short; sheathing leaf bases form a false stem to 40cm; leaves elliptic-lanceolate, 15–40 × 5–10cm, appearing with the flowers; scape derived from the centre of the false stem, to 65cm, inflorescence subglobose, many-flowered; flowers pale red; pedicels c. 2cm. Forest; 1931m.
Distr.: Nigeria to Gabon [lower Guinea].
IUCN: NT
Dom: Cheek sight record 30 9/2006.

Anthericaceae

M. Cheek (K)

Chlorophytum comosum (Thunb.) Jacq. var. **sparsiflorum** (Baker) A.D.Poulsen & Nordal
Bot. J. Linn. Soc. 148(1): 15 (2005).
Herb, 25–60cm, drying light green, sometimes viviparous; leaves oblanceolate or oblanceolate-ligulate, c. 25 × 6cm, acute-mucronate, base tapering into a variably defined petiole; inflorescence about as long as leaves, or longer. Forest; 1600m.
Distr.: Sierra Leone to Kenya [afromontane].
IUCN: LC
Dom: Cheek 13452 fr., 9/2006.
Local name: Nken (Cheek 13452).

Araceae

A. Haigh (K) & P.C. Boyce

Fl. Cameroun 31 (1988).

Amorphophallus staudtii (Engl.) N.E.Br.
Herb; leaf like a tattered umbrella, leaflets acute, not fishtail shaped; petiole to 1.2m, smooth; inflorescence pale dirty cream-white; peduncle very short; spadix base not swollen. Forest; 1590m.
Distr.: Cameroon, Equatorial Guinea [lower Guinea].
IUCN: NT
Dom: Dom photographic record 1 2/2006.
Note: known from only 10 sites in Cameroon, where it is likely threatened in the Bamenda Highlands by agricultural encroachment into existing forest patches. Also recorded from Equatorial Guinea, but poorly documented here.

Amorphophallus sp. cf. **zenkeri** (Engl.) N.E.Br.
Leaf, 25cm tall, single; petiole erect, cylindric, smooth without prickles; blade pedate of 7 leaflets; leaflets elliptic, c. 10 × 4cm, acumen 1cm, lateral nerves 8 pairs; inflorescence and fruit unknown. Forest, near stream; 1600m.
Dom: Cheek 13461 9/2006.
Note: flowers needed to identify firmly (Cheek 13461).

Anchomanes difformis (Blume) Engl.
Syn. Anchomanes difformis (Blume) Engl. var. **pallidus** (Hook.) Hepper
Syn. Anchomanes welwitschii Rendle
Herb; leaf like a tattered umbrella, leaflets fishtail-shaped; petiole spiny (prickles); spathe green tinged purple; styles straight but very short, smooth, green. Forest & forest margins; 1600–1633m.
Distr.: [Guineo-Congolian].
Dom: Dom photographic record 2 2/2006; Cheek sight record 10 9/2006.
Note: A. hookeri was until recently treated as a synonym of A. difformis but the two are separable on differences in the styles. However, the distributions of these taxa have not been fully defined as yet; no conservation assessments can be made until this is clarified.

Culcasia ekongoloi Ntepe-Nyame
Fl. Cameroun 31: 87 (1988).
Slender climber; stems minutely roughened; leaves with transparent lines; petiole as long as lamina. Stream banks in forest; 1600m.
Distr.: Nigeria to Congo (Kinshasa) & CAR [Guineo-Congolian].
IUCN: LC
Dom: Cheek 13457 fl., 9/2006.

Commelinaceae

M. Cheek (K) & J.-M. Onana (YA)

Aneilema beniniense (P.Beauv.) Kunth
Weak erect herb, to c. 1m; leaves elliptic, c. 15 × 5cm, acuminate, sessile or shortly petiolate; inflorescence terminal, dense, c. 3–4 × 3–4cm; flowers white or mauve; fruits longer than broad, apex rounded. Lowland farmbush, forest, stream banks & Aframomum thicket; 1600m.
Distr.: tropical Africa [tropical Africa].
IUCN: LC
Dom: Cheek 13458 fl., 9/2006.
Local name: Leh (Cheek 13458).

Commelina africana L. var. *africana*
Prostrate herb with rooting stems, to 1m long; leaves ovate-elliptic, c. 4 × 2cm, apex acute; inflorescensces subtended by spathes; flowers yellow, open 7–10 am. Montane grassland; 1659m.
Distr.: Guinea (Bissau) to South Africa [afromontane].
IUCN: LC
Dom: Cheek 13531 fl., 9/2006.
Local name: Ee Wor (Cheek 13531). **Uses:** MEDICINES (Cheek 13531).

Commelina benghalensis L. var. *hirsuta* C.B.Clarke
Erect herb, c. 30(–150)cm tall; leaves ovate, to 6 × 3.5cm, subacuminate, truncate, petiolate; sheath with conspicuous rusty hairs all over outside; spathe c. 2 × 1cm; flowers bright blue, open 8.30–12am. Lower montane to submontane farmbush; 1711m.
Distr.: Guinea (Conakry) to Malawi [afromontane].
IUCN: LC
Dom: Cheek 13537 fl., 9/2006.
Uses: MEDICINES (Cheek 13537).
Note: for Cheek 13537, the hairs on the sheath should be brown not white, but perhaps blanched by alcohol? (Cheek, xi.2008).

Cyanotis barbata D.Don
Erect herb, 10–30cm; with underground rootstock; leaves linear-lanceolate, to 12 × 1cm, white pubescent, sessile; spathes 4–5, c. 1 × 0.5cm, pedunculate; flowers blue, actinomorphic; filaments bearded. Montane grassland; 1797m.
Distr.: tropical Africa & the Himalayas [palaeotropics (montane)].
IUCN: LC
Dom: Cheek 13561 fl., 9/2006; 13569 fl., 9/2006.
Note: Cheek 13569 appears the same as Cheek 13561 despite the differing flowering hour and much taller, lanky habit (inflorescence not dissected however) (Cheek, xi.2008).

Cyanotis longifolia Benth. var. *longifolia*
Erect, unbranched herb, 0.5m, from underground rootstock; densely cotton wool-hairy; internodes 5-12cm; leaf-blade strap-like, 10-12 × 0.6cm, erect, sheath 4cm; inflorescence a terminal head 2.5cm; spathes densely packed, 15-25, c. 1cm long; flowers pale blue; petals 3. Grassland; 1931m.
Distr.: widespread in savanna of tropical Africa [tropical Africa].
IUCN: LC
Dom: Cheek 13570 9/2006.
Note: in northern, drier parts of the Cameroon Highlands (Cheek, xi.2008).

Palisota mannii C.B.Clarke
Herb, 20-80cm; lacking aerial stem; leaves forming a basal rosette, lanceolate or lanceolate-obovate, 25-40 × 5-9cm, apex acuminate, base cuneate, margin hairy, lower surface white, glabrous; inflorescence cylindrical, c. 12-18 × 3.5cm; peduncle 10-50cm long; pedicels > flowers; rarely bracteose; flowers white; fruits red. Forest & forest-grassland transition; 1600m.
Distr.: S Nigeria to Uganda [lower Guinea & Congolian].
IUCN: LC
Dom: Onana 3622 2/2006.
Note: R. Faden is currently revising the *P. mannii* species complex.

Cyperaceae

A.M. Muasya (BOL), K. Hoenselaar (K) & M. Cheek (K)

Ascolepis protea Welw. subsp. *protea*
Annual or perennial herb; stems slender, c. 0.5mm diam., often thickened at the base with pale fibrous sheaths; inflorescence consisting of a single spikelet with white, often curved, scales in a radiate head about 6-8mm across. Swamps, wet grassland, granite outcrops; 1931m.
Distr.: Senegal to Ethiopia and S to Zimbabwe [tropical Africa].
IUCN: LC
Dom: Cheek 13583 9/2006.

Bulbostylis densa (Wall.) Hand.-Mazz. var. *cameroonensis* Hooper
Annual herb, to ± 30cm; stem deeply grooved, 0.2–0.4mm thick; leaves canaliculate, grooved, 0.2–0.3mm broad; inflorescence usually a compact umbel, somewhat contracted, with 3–8 shortly pedicellate spikelets, each one 2–5 × 1.5–3mm; glumes few, each standing out from its neighbour, dark brown with conspicuous pale green or grey midrib. Grassland; 1797m.
Distr.: Mt Cameroon and Bamenda Highlands [W Cameroon Uplands].
IUCN: VU
Dom: Cheek 13552 fl., 9/2006.

Bulbostylis densa (Wall.) Hand.-Mazz. var. *densa*
Annual herb, very variable in size; spikelets ovate to broadly elliptic, always pedicillate; glumes broadly ovate, clearly keeled. Grassland, rocky slopes; 1931m.
Distr.: palaeotropical [montane].
IUCN: LC
Dom: Cheek 13582 fr., 9/2006.
Note: herb, to 25cm tall, similar to var. *cameroonensis* (Cheek 13552) but tall, with spikelet differences (Cheek, vi.2009).

Coleochloa abyssinica (Hochst ex A.Rich) Gilly var. *abyssinica*
F.T.A. 8: 512 (1902); Brittonia 5: 14 (1943).
Robust plant, with branching scaly stolons; stems 40–80 × 0.1–0.4cm; leaf-blades to 30 × 0.2–0.7cm, folded; ligule 1–2mm; inflorescence a diffuse panicle, 2–6-branched from upper leaf-sheaths; spikes 5–8 × c. 3mm, of numerous densely clustered spikelets; glumes pale or dark brown or red brown, glossy, glabrous. In rock crevices, shallow soils over rocks, grasslands, often epiphytic on trees in montane forest; 1931m.
Distr.: NE Nigeria, W Cameroon, E & NE tropical Africa, Angola [afromontane].
IUCN: LC
Dom: Cheek 13585 fl., 9/2006.

Coleochloa domensis Muasya & D.A. Simpson sp. nov. ined.
Kew Bull. (2010) (in press)
Epiphytic herb, 30cm, erect, tufted; culms c. 6-leaved; leaf base sheath 2-4cm, apex long-hairy; blade 25 × 0.4cm; inflorescence spike-like, 6-8 × 1cm, brown; flowers 3-5-clustered, c. 7mm long, erect, cluster stalks to 1cm; styles long exserted by 3mm, black, recurved, conspicuous. Forest; 1600–1831m.
Distr.: known only from Dom [Cameroon endemic].
Dom: Cheek 13438 9/2006; 13647 9/2006.
Local name: Njing-Njing (Cheek 13438).

Cyperus dilatatus Schumach. & Thonn.
Erect herb, 0.5m, glabrous; culms in clumps; leaves 4; sheaths 4-5cm; blade c. 15cm × 2mm; inflorescence umbellate; bracts leafy, 3, unequal, longest 20cm × 3mm; spike clusters c. 8, stalks 0-8cm; spikes 2-6 per cluster, each cylindric to 7 × 1.5cm, scales concave, brown, midrib 2-ribbed, acumen hairy. Grassland; 1931m.
Distr.: [tropical Africa].
IUCN: LC
Dom: Cheek 13601 fl., 9/2006.
Local name: Njin-Jing (Cheek 13601).

Kyllinga triceps Rottb.
Syn. *Kyllinga tenuifolia* Steud.
Erect herb, 20cm, citrus scented, glabrous; leaves 2 per culm, 10-15cm, 2mm wide, erect; spike dull white, single, erect, shortly cylindrical, 8 × 5mm, dull; spikelets ovoid, 1.5mm, sessile, patent, longitudinally ridged; bracts 3, leafy, patent, all unequal, shortest c. 3cm, longest c. 8cm. Submontane grassland; 1633m.
Distr.: [amphi-Atlantic].
IUCN: LC
Dom: Cheek 13514 fl., 9/2006.
Local name: Nsaan-Sah (Cheek 13514).

Pycreus atrorubidus Nelmes
Herb, 10-20cm tall, annual, glabrous; leaves 6-8 per culm, all basal, 5-8cm × 0.5-1mm; spike cluster appears lateral, 4cm below apex of inflorescence axis, and a suberect bract 2cm; spike 10 per cluster, flattened, ovate, 5 × 2mm, acute; scales 2-ranked, brown-black, with yellow margins. Common on rocks; 1797m.
Distr.: Guinea (Conakry), Sierra Leone, Liberia, Ivory Coast, W Cameroon and SE Africa [afromontane].
IUCN: LC
Dom: Cheek 13560 fl., 9/2006.

Scleria interrupta Rich.
Adansonia (ser. 2) 16: 216 (1976).
Syn. *Scleria hirtella* sensu Napper
Erect perennial herb, 10-15cm; culms densely clumped, 3-angled, bearing 3 cauline leaves subtending sheath; leaves 11 × 0.2cm, scattered with long white hairs; inflorescence rachis c. 5cm, being 10-15 evenly spread spikes; spikes 3-4mm, loose, open; glumes red brown, hairs as leaves. Rocky slopes; 1931m.
Distr.: [amphi-Atlantic].
IUCN: LC
Dom: Cheek 13581 fl., 9/2006.

Scleria melanotricha Hochst. ex A.Rich. var. grata (Nelmes) Lye
Haines R. & Lye K., The Sedges and Rushes of East Africa: 339 (1983).
A slender annual; stems 10-50cm; leaves 1-2mm wide, hairy; inflorescence spicate, 3-20cm; glomerules paired, shortly pedunculate; glumes straw-coloured to reddish brown, densely hairy. Wet places; 1797m.
Distr.: Sierra Leone, Ivory Coast, N Nigeria, W Cameroon, scattered from Ethiopia to Zambia [montane].
IUCN: LC
Dom: Cheek 13559 fl., 9/2006.

Dioscoreaceae

P. Wilkin (K)

Dioscorea bulbifera L.
Herbaceous climber, glabrous, to 3-7m; stems left-twining (sinistrorse); leaves alternate with a pair of membranous

semicircular lateral projections clasping stem at petiole base, apex short-acuminate, not thickened. Farmbush; 1600m.
Distr.: Senegal, Guinea (Conakry), Sierra Leone, Ivory Coast, Burkina Faso, Ghana, Nigeria, Cameroon [palaeotropics].
IUCN: LC
Dom: Cheek 13435 9/2006.
Note: not eaten (Cheek 13435).

Dioscorea preussii Pax subsp. *preussii*
Robust non-spiny climber, to 10m; stems often 6-winged, subglabrous, with a few ± caducous, medifixed (T-shaped) hairs, also on the inflorescence and leaf apices; leaves alternate, broadly ovate, obliquely acuminate, deeply cordate, 10–30 × 8–35cm, villous-tomentose beneath. Forest & farmbush; 1600m.
Distr.: Senegal to Uganda, Angola, Mozambique [tropical Africa].
IUCN: LC
Dom: Cheek 13430 9/2006.

Dioscorea schimperiana Hochst. ex Kunth
Climber, 3-7m, twining, unarmed, pubescent; leaves subopposite, ovate, 10-15 × 10-15cm, acuminate, deeply cordate at the base, petiolate; male and female flowers on long fascicled spikes; capsule c. 4cm diam., glabrescent, 3 acute orbicular lobes and an axis c. 2.5cm long; seeds orbicular, winged. Submontane forest; 1600–1607m.
Distr.: Nigeria to Ethiopia, southwards to Malawi and Angola [tropical Africa].
IUCN: LC
Dom: Cheek 13403 fl., 9/2006; 13415 fr., 9/2006.
Note: not eaten (tubers hurt mouth if eaten) (Cheek 13415).

Dracaenaceae

M. Cheek (K)

Dracaena arborea Link
Tree, 10–20m, trunk 30cm, with aerial roots, several-branched; leaves in dense heads, sword-shaped, 50–120 × 4.5–6cm, widest above the middle, apex acute, mucro to 3mm, base clasping stem for 3/4 circumference; inflorescence pendulous, to 1.5m; perianth cream-white, c. 1.5cm; fruit to 2cm, orange-red. Submontane forest & planted; 1931m.
Distr.: Sierra Leone to Angola [Guineo-Congolian].
IUCN: LC
Dom: Cheek sight record 28 9/2006.

Dracaena fragrans (L.) Ker-Gawl.
Syn. *Dracaena deisteliana* Engl.
Herb, to 3m, few-stemmed; stalk 1cm diam.; leaves sword-shaped, to 70 × 9cm; inflorescence terminal, erect; flowers white with pink lines, very fragrant. Forest; 1831m.
Distr.: tropical Africa [afromontane].
IUCN: LC
Dom: Cheek sight record 38 9/2006.

Dracaena laxissima Engl.
Shrub, 1.5m, glabrous; stems 3-4mm diam., pale brown, striate; internodes 2-3cm; leaves evenly spread, alternate, elliptic, c. 15 × 5cm, acumen 1cm with a 5mm filament tip, base acute, lateral nerves c. 30 pairs, parallel; petiole 1cm, well-defined, base 9/10 amplexical; inflorescence terminal, paniculate, 15cm, diffuse, pendant, with c. 6 patent lateral branches each c. 5cm long. Forest; 1600m.
Distr.: Nigeria to E Africa, Sudan, Zambia and Malawi [tropical Africa].
IUCN: LC

Dom: <u>Cheek 13431</u> 9/2006.

Gramineae

T. Cope (K)

Andropogon schirensis A.Rich.
Syn. *Andropogon dummeri* Stapf
Erect caespitose perennial, to 2m; leaf blades linear, to 45 × 1.4cm, mostly cauline; racemes 6-12cm; sessile spikelets, 5-7mm. Deciduous bushland and wooded grassland; 1931m.
Distr.: tropical and subtropical Africa [afromontane].
IUCN: LC
Dom: <u>Cheek 13580</u> fl., 9/2006.

Ctenium ledermannii Pilg.
Perennial, 60-90cm; 2-5, dark green, paired (rarely solitary) digitate spikes, 10-14cm. Forest edge; 1931m.
Distr.: Nigeria to W Cameroon [afromontane].
IUCN: NT
Dom: <u>Cheek 13591</u> 9/2006.

Eragrostis pobeguinii C.E.Hubb.
Densely caespitose perennial, about 30cm; panicle scantily branched, bearing to 15 spikelets, pallid to olive-grey; leaf-blades to 2mm wide, usually rolled and setaceous; basal sheaths bulbously swollen and hardened below. Grassland; 1931m.
Distr.: Senegal, Guinea (Conakry), Ghana, Cameroon [upper & lower Guinea].
IUCN: LC
Dom: <u>Cheek 13596</u> fl., 9/2006.

Eragrostis tenuifolia (A.Rich.) Hochst. ex Steud.
Perennial, 35–70cm, caespitose, glabrous; basal sheaths laterally compressed; leaf-blades 5–20cm × 1–4mm, ribbon-like, attenuate; ligule hyaline with a row of white hairs, 2mm; inflorescence an open elliptic panicle, 5–15cm, with long feathery branches; spikelets on long pedicels, 4–11mm; white hairs at axils of branches; spikelets 7–15 × 1.5mm, 4–15 florets, linear, dark green, margins conspicuously saw-toothed; lower glume about 2/3 length of upper, the latter 0.8–1.4mm long and barely reaching base of adjacent lemma; paleas persisting long after lemmas have fallen. Wasteland, fallow, grassland; 1931m.
Distr.: [pantropical].
IUCN: LC
Dom: <u>Cheek 13603</u> 9/2006.
Local name: Nsuleh (Cheek 13603).

Loudetia simplex (Nees) C.E.Hubb.
Syn. *Loudetia camerunensis* (Stapf) C.E.Hubb.
Perennial; culms caespitose, erect, up to 1.5m; nodes glabrous or having a ring of hairs; basal sheaths decaying into fibres, pubescent, sometimes glabrous; leaf-blades linear, 10–30cm × 2–5mm; inflorescence a lax panicle, linear–ovate, 10–30cm, branches occasionally verticillate; spikelets 8–14mm, narrowly lanceolate, brown, glabrous; upper lemma acutely 2–lobed, the lobes 0.3–1mm long; lower glume obtuse, usually 1/3 length of spikelet; awns 2.5–5cm; stamens 2. Grassland, stony soils, marshes, swamps; 1931m.
Distr.: tropical and S Africa [afromontane].
IUCN: LC
Dom: <u>Cheek 13577</u> 9/2006.

Panicum acrotrichum Hook.f.
Straggling perennial, rooting from the nodes; leaf-blades lanceolate, 3-8 × 0.8-2.3cm, with transverse veins; panicle to 11cm, branched. Forest; 1810m.

Distr.: W Cameroon & Bioko [W Cameroon Uplands (montane)].
IUCN: VU
Dom: <u>Cheek 13659</u> fl., 9/2006.
Note: Cheek 13659 is tentatively placed in this taxon by Cope, xi.2008 (badly smutted).

Pennisetum clandestinum Hochst. ex Chiov.
Perennial, c. 4cm tall; rhizomatous and stoloniferous, forming a dense mat; lower leaf-blades distichous, sheaths flattened laterally; inflorescence within an involucre of fine hairs, each with 2–4 subsessile spikelets enclosed in the uppermost leaf sheath; spikelets 10–20mm, narrowly lanceolate; stigmas white, 4cm; stamens 3–5cm with silvery filaments conspicuously exserted from leaf–sheath when in flower; anthers light brown, 4.5mm. Meadows and savanna at high altitude; 1931m.
Distr.: E Africa; introduced into Cameroon as fodder and to combat erosion [afromontane].
IUCN: LC
Dom: <u>Cheek 13574</u> 9/2006.

Setaria megaphylla (Steud.) T.Durand & Schinz
Syn. *Setaria chevalieri* Stapf
Perennial, 1–3m, rhizomatous; culms 5–10mm diam. at base; leaf-blades linear to lanceolate, 30–70cm × 1–10cm; inflorescence a linear-lanceolate panicle, length 20–50cm; branches 4–15cm, straight, rigid; spikelets subtended by bristle(s), 2.5–3.5mm, ovate to elliptic; lower lemma equalling or very shortly exceeding upper lemma; upper lemma smooth, shiny; palea of lower floret as long as the lemma. Forested areas near roads; forest edges; 1607m.
Distr.: tropical Africa, tropical America [amphi-Atlantic].
IUCN: LC
Dom: <u>Cheek sight record 9</u> 9/2006.

Setaria sphacelata (Schumach.) Stapf & C.E.Hubbard ex M.B.Moss var. *sericea* (Stapf) W.D.Clayton
F.T.E.A. Gramineae: 528 (1982); Agric. Univ. Wag. Papers 92(1): 297 (1992).
Syn. *Setaria anceps* Stapf ex Massey
Tufted perennial arising from short rhizomes; culms 4-10-noded, 3-6mm diam., up to 2m high; basal leaf-sheaths often conspicuously flabellate; leaf blades 3-10mm wide, glabrous; panicle spiciform, 7-25cm long; spikelets pallid to purple with fulvous bristles, elliptic, oblique; lower glume up to 1/2 as long as the spikelet. Submontane grassland; 1633m.
Distr.: tropical Africa and South Africa [tropical Africa and South Africa].
IUCN: LC
Dom: <u>Cheek 13505</u> fl., 9/2006.

Sporobolus africanus (Poir.) Robyns & Tournay
Syn. *Sporobolus indicus* (L.) R.Br. var. *capensis* Engl.
Agric. Univ. Wag. Papers 92(1): 157 (1992).
Perennial, 60cm, densely caespitose; leaf-blades rigid, 10–35cm × 3–4mm; ligule ciliate; inflorescence a linear panicle, 6–25cm, narrow to subspiciform, branches appressed and mostly 1–2cm long; spikelets 1.8–2.1mm, not acuminate, olive green; upper glume up to 2/3 length of spikelet. Humid montane areas, swamps; 1931m.
Distr.: tropical and subtropical Africa, Sri Lanka, Philippines, Australia, New Zealand [montane].
IUCN: LC
Dom: <u>Cheek 13579</u> 9/2006.
Local name: Fiuw (Cheek 13579).

Sporobolus festivus Hochst. ex A.Rich.
Densely caespitose perennial; culms 10-60cm high, slender; leaf blades convolute, 2-7 × 0.1-0.2cm; panicle narrowly ovate, 3-22cm; branches tinged with red; spikelets small, 1-1.5mm long, grey-purplish; grain ellipsoid to obovoid, 0.4-0.7mm. Rocky outcrops; 1797m.
Distr.: Mauritania, E to Somalia, and S to South Africa [subsaharan Africa].
IUCN: LC
Dom: Cheek 13556 9/2006.

Trichopteryx elegantula (Hook.f.) Stapf
Annual, 5–20(–30)cm; gregarious; culms erect or geniculately ascending; leaf-blades 1–2cm × 2–4mm, lanceolate, pilose; ligule a row of hairs 1–1.5mm; inflorescence a panicle, 3–10cm, obovate, lax, scarcely extending from sheath; branches filiform, ascending; spikelets solitary or in pairs; pedicels capillary, 4–8mm; spikelets 2.5–3.5mm, lanceolate, pilose or glabrous; glumes glabrous or bearing stiff tubercle-based hairs; main awn 10–13mm, caudate, base black, upper part green; lateral awns 4mm. Slopes of hills, between rocks; 1633–1931m.
Distr.: Sierra Leone, Nigeria, Cameroon, Congo (Kinshasa), Rwanda, Burundi, E Africa [afromontane].
IUCN: LC
Dom: Cheek 13487 9/2006; 13608 fl., 9/2006.

Tripogon major Hook.f.
Densely caespitose perennial, about 30cm; leaf-blades narrow, involute, to 15cm; spikes 8-25cm long, erect; spikelets 6-18-flowered, elliptic to narrowly oblong, 13-25mm long. Grassland and forest-grassland transition; 1931m.
Distr.: tropical Africa [afromontane].
IUCN: LC
Dom: Cheek 13578 9/2006.

Hyacinthaceae

M. Cheek (K)

Drimia altissima (L.f.) Ker Gawl.
Bothalia 13: 452 (1981).
Syn. Urginea altissima (L.f.) Baker
Herb, to 2m; bulb globose, 10-15cm diam.; leaves 5-6, lorate-lanceolate, glabrous, 30-45cm long; peduncle to 2m; raceme cylindrical, 30-60cm long; perianth campanulate, c. 7mm long; capsule globose, c. 1cm diam.; seeds black. Grassland; 1815m.
Distr.: Sierra Leone down to South Africa [tropical & subtropical Africa].
IUCN: LC
Dom: Harvey 317 2/2006.

Iridaceae

M. Cheek (K)

Gladiolus aequinoctialis Herb.
Fl. Zamb. 12(4): 102 (1993); F.T.E.A. Iridaceae: 84 (1996); Goldblatt P., Gladiolus in Tropical Africa: 287 (1996).
Syn. Acidanthera aequinoctialis (Herb.) Baker
Herb, (40–)90–120cm; corm 2–3cm across; leaves 4–10, lower 2–5 basal, lanceolate, (0.6–)1.2–1.7cm across; spike (3–)5-8-flowered; bracts 5–8cm long; flowers white, showy, the lower 3 tepals streaked purple; perianth tube cylindric, (8.5–)12–14cm; tepals lanceolate, ± equal, 3.5–4 × 1.8–

2.0cm; capsules 1.8–2.0cm long. Rocky places, often on wet ledges on steep cliffs & stony grassland; 1931m.
Distr.: Sierra Leone, W Cameroon & Bioko [upper & lower Guinea].
IUCN: NT
Dom: Cheek 13584 fl., 9/2006.
Local name: Ntu-Bun (Cheek 13584).
Note: here we follow Goldblatt (1996), who in his mongraph of the tropical African species of Gladiolus does not recognise varieties under this name.

Gladiolus dalenii van Geel subsp. andongensis (Welw. ex Baker) Goldblatt ex Geerinck
Flora Zambesiaca 12(4): 93 (1993); .Taxonomania 6: 12 (2002).
Herb, to 60–90cm, often puberulous; corm (15–)20–30mm diam.; tunics of brittle membranous layers, sometimes fibrous, reddish-brown, usually with numerous tiny cormlets around the base; leaves not contemporaneous with the flowers, appearing on separate shoots after flowering, at least 2, narrowly lanceolate, 30–50cm × 4–16mm; spike (2–)3–9-flowered; flowers orange or rarely yellow; perianth tube 25–33(–40)mm, the dorsal sepal exceeding the laterals. Grassland, savanna, woodland.
Distr.: Guinea (Conakry), Sierra Leone, Ivory Coast, Cameroon, CAR, Congo (Kinshasa) to Ethiopia, E Africa, S to Mozambique, Zimbabwe [tropical Africa].
IUCN: LC
Dom: Corcoran sight record 2 4/2005.

Musaceae

M. Cheek (K)

Fl. Cameroun 4 (1965).

Ensete gilletii (De Wild.) Cheesman
Monocarpic herb, 1.5-3m; leaves spread across the stem and not aggregated towards the apex, lower leaves to 1.5m, reducing in size upwards so that the upper leaves become bracteate; male bracts 4.5-9 × 17-25cm; stamens 6-12mm; female flower with rudimentary stamens; fruit squat, angular, rather obconic, about 5cm; seeds 7-9mm diam., brown or black, hard. A wild banana of hilly grassland savanna; 1633m.
Distr.: Guinea (Conakry), Mali, Sierra Leone, Ivory Coast, Ghana, Togo, Benin, Nigeria, Cameroon, CAR, Congo (Kinshasa), Malawi, Angola [tropical Africa].
IUCN: LC
Dom: Cheek sight record 17 9/2006.
Local name: Gom Ten (Cheek sight record 17).

Orchidaceae

D. Roberts (K), P. Cribb (K) & J.-M. Onana (YA)

Fl. Cameroun 34 (1998), 35 (2001) & 36 (2001).

Bulbophyllum calvum Summerh.
Epiphytic herb; pseudobulbs 2-3cm diam.; sepals pale yellow-green; labellum red with yellow base and white flecks above. Forest, woodland; 1831m.
Distr.: Mt. Cameroon and E Nigeria [W Cameroon Uplands (montane)].
IUCN: NT
Dom: Cheek 13616 fl., 9/2006.
Note: tentative determination. Owing to the complexities of acquiring CITES export permits from Cameroon, subsequent

import permits to the UK, shipping the permits back to Cameroon, and finally shipping the specimens and permits to the UK, within a six month time-frame, the above orchid has not made it to a specialist in time for naming prior to the publication of the checklist.

Bulbophyllum cochleatum Lindl. var. cochleatum

Epiphyte; pseudobulbs bifoliate, 0.8–7cm apart, 1.5–11 × 0.4–1.3cm; leaves 2.8–23 × 0.3–1.8cm; inflorescence 8–55cm; peduncle 5.2–43cm; rachis not thickened, 2.8–12cm, 14–64(–84)-flowered; sepals and petals green, often stained purple-red or entirely so, with a green base; labellum dark purple red, occasionally with a yellow centre, with marginal hairs ≥ labellum width. Forest; woodland; 1797–1831m.
Distr.: Guinea to Bioko, São Tomé, Cameroon, Gabon, Sudan to E and S Africa [afromontane].
IUCN: LC
Dom: Cheek 13615 9/2006; 13663 fl., 9/2006.
Note: Cheek 13615 and 13663 have tentatively been placed in this taxon, see CITES note after *Bulbophyllum calvum*.

Bulbophyllum nr. cochleatum Lindl.

Epiphyte; pseudobulb short, slender; flowers scarlet. Forest patch, abundant on one *Gnidia glauca*, not seen elsewhere; 1797m.
Dom: Cheek 13664 fl., 9/2006.
Note: tentative determination. See CITES note after *Bulbophyllum calvum*.

Bulbophyllum maximum (Lindl.) Rchb.f.

Syn. *Bulbophyllum oxypterum* (Lindl.) Rchb.f.
Epiphyte; pseudobulbs 2(–3)-leafed, 2–10cm apart, 3.5–10 × 1–3cm; leaves oblong to linear-lanceolate, maximum width usually just above middle, 3.8–20 × 1.3–5cm; inflorescence 15–90cm; 16–120-flowered; rachis bladelike, 6–56 × 8–50mm; floral bracts spreading to reflexed, 2.5–7 × 2–4mm; flowers yellowish or greenish, spotted purple. Lowland primary and secondary forest, montane forest, savanna wooodland; 1490m.
Distr.: [tropical & subtropical Africa].
IUCN: LC
Dom: Dom photographic record 26 2/2006.

Bulbophyllum schimperianum Kraenzl.

Epiphyte; pseudobulbs tightly clustered, unifoliate, ovoid or conical, 1.2–2.5 × 0.8–1.7cm, yellow; leaf coriaceous, oblong or oblong-lanceolate, slightly falcate, 4.5–13.5 × 1.2–2.5cm, minutely apiculate; inflorescence 10–33cm, densely many-flowered; rachis terete; flowers white or cream, ± flushed pink. Forest; 1810m.
Distr.: Liberia, Nigeria, Cameroon, Gabon, Congo (Brazzaville), Congo (Kinshasa), CAR, Uganda [Guineo-Congolian].
IUCN: LC
Dom: Cheek 13658 fl., 9/2006.
Note: Cheek 13658 is tentatively placed in this taxon, see CITES note after *Bulbophyllum calvum*.

Calyptrochilum christyanum (Rchb.f.) Summerh.

Epiphyte; stem woody, to 50cm, pendulous or almost horizontal; inflorescence usually 6-9 or sometimes to 12-flowered, rather lax, to 4cm; flowers white or cream, the base of the labellum often greenish, yellow or orange. Forest, woodland; 1600m.
Distr.: tropical Africa [tropical Africa].
IUCN: LC
Dom: Onana 3624 2/2006.

Diaphananthe bueae (Schltr.) Schltr.

Epiphyte; flowers white and green; sepals 6-8.5mm. Forest; 1831m.
Distr.: Cameroon, Ivory Coast and Uganda [lower Guinea (montane)].
IUCN: EN
Dom: Cheek 13617 fl., 9/2006.
Note: Cheek 13617 is tentatively placed in this taxon, better material needed. See CITES note after *Bulbophyllum calvum*.

Eulophia odontoglossa Rchb.f.

Linnea 19: 373 (1846); Adansonia (Sér. 2) 8: 292 (1968); F.T.E.A. Orchidaceae: 458 (1989); Fl. Afr. Cent. Orchidaceae: 680 (1992).
Syn. *Eulophia shupangae* (Rchb.f.) Kraenzl.
Terrestrial herb, 0.6–1.0m; leaves 5–6, erect, plicate, oblanceolate, acuminate, 40–70 × 1–2.1cm, basal 3 sheathing; inflorescence densely many-flowered; flowers yellow; labellum with yellow, orange or red papillae. Grassland, bushland, rocky areas.
Distr.: tropical and subtropical Africa [afromontane].
IUCN: LC
Dom: Corcoran sight record 3 4/2005.

Habenaria malacophylla Rchb.f. var. malacophylla

A slender terrestrial herb, 0.3-1m; stem leafy in centre part, bare below; inflorescence a long loose raceme, 8–34 × 2.5–3.5cm; bracts lanceolate, 0.9–2.0cm; flowers numerous, small, green; petals bipartite near the base, anterior lobe 5–9.5 × 0..3–0.8mm, posterior lobe 4–7.3 × 0.4–1.5mm; labellum trilobed, median lobe 4.5–8 × 0.8–1mm; spur 9–18mm. Woodland and grassland; 1600–1810m.
Distr.: Sierra Leone, Nigeria, Cameroon, Ethiopia, E Africa, Malawi, Zambia, Zimbabwe, South Africa, Oman [afromontane & Arabia (montane)].
IUCN: LC
Dom: Cheek 13406 fl., 9/2006; 13654 fl., 9/2006.

Habenaria mannii Hook.f.

Terrestrial herb, to 70cm; stem leafy; leaves linear-lanceolate, 5–15 × 0.5–2.0cm; inflorescence a dense raceme, 6–22 × 3–4.5cm, of few–25 flowers; flowers pale green; labellum tripartite, easily recognised by outside of lateral lobes being conspicuously pectinate; spur 11–14(–28)mm. Grassland; 1780m.
Distr.: Nigeria, Bioko, Cameroon [W Cameroon Uplands].
IUCN: NT
Dom: Cheek 13540 9/2006.
Note: tentative determination. See CITES note after *Bulbophyllum calvum*.

Habenaria nigrescens Summerh.

Terrestrial herb, 15–70cm; leaves 9, arranged spirally, sheathing basally, linear to linear-lanceolate, 7–13 × 0.5–1.0cm; inflorescence spicate, lax to dense, 5–13cm, 4–17-flowered; floral bracts 10–15mm; flowers pale green or yellow-green; labellum deeply trilobed, median lobe 4.3–4.5 × 1.2mm; spur 6–8.5mm. Submontane grassland; 1931m.
Distr.: SE Nigeria, W Cameroon [W Cameroon Uplands].
IUCN: VU
Dom: Cheek 13592 fl., 9/2006.
Note: tentative determination, see CITES note after *Bulbophyllum calvum*.

Polystachya alpina Lindl.

Epiphyte to 20cm; stems swollen at base; leaves 2.5 × 0.3–1.2cm; inflorescence lax, 1.5–6(–9)cm, 7–20-flowered; peduncle and rachis densely pubescent; flowers small, non-resupinate; tepals a sparkling 'crystalline-white', tinged pink;

labellum 7–9 × 4m; spur 6mm, cylindrical-sacciform. Forest or woodland; 1797m.
Distr.: SE Nigeria, W Cameroon and Bioko [W Cameroon Uplands].
IUCN: NT
Dom: Cheek 13665 fl., 9/2006.
Note: tentative determination. See CITES note after *Bulbophyllum calvum*.

Tridactyle tridactylites (Rolfe) Schltr.
Epiphyte, or epilith; stems pendent 0.4 to 1.6m, robust; leaves numerous, 6–21 × 0.6–1.3cm, linear or linear-lanceolate; inflorescence lax, 1.7–10cm, to 18-flowered; flowers small, resupinate, yellow, orange or brownish-orange, sometimes tinged green, fragrant; labellum 4–5 × 6.5mm, side lobes ± equal in length to mid-lobe, entire or rarely slightly bifid; spur 6–11mm. Forest; 1490m.
Distr.: Sierra Leone to Congo (Kinshasa), E Africa, S to Mozambique, Malawi, Zambia, Angola, Zimbabwe [afromontane].
IUCN: LC
Dom: Harvey 327 2/2006.

Palmae

W.J. Baker (K)

Phoenix reclinata L.
Tree, or shrub; stems clustered, fruiting when 1–10m tall; leaves pinnately compound, leaflet apices often spine-like; fruits ellipsoid, fleshy, 2 × 1cm, orange, ripening brown. Forest, farms and farmbush; 1570m.
Distr.: tropical and subtropical Africa [subsaharan Africa].
IUCN: LC
Dom: Dom photographic record 10 2/2006.

Raphia mambillensis Otedoh
J. Niger. Inst. Oil Palm Res. 6(22): 163 (1982).
Palm with no clear trunk; stem very short, covered by petioles; fronds clustered at ground level. Common near villages at the edge of water courses; 1633m.
Distr.: Nigeria, Cameroon, CAR, Sudan.
IUCN: NT
Dom: Cheek sight record 11 9/2006.
Local name: Ki Keih (Cheek sight record 11).

Smilacaceae

M. Cheek (K)

Smilax anceps Willd.
Meded. Land. Wag. 82(3): 219 (1982).
Syn. *Smilax kraussiana* Meisn.
Climber, to 7m; stem spiny; leaves coriaceous, alternate, elliptic, c. 14 × 8cm, mucron 0.5cm, base obtuse, nerve palmate, 3–5; petiole c. 2cm; inflorescence terminal, umbellate, 5cm diam. Forest; 1640m.
Distr.: Senegal to S Africa [tropical & subtropical Africa].
IUCN: LC
Dom: Nana 142 4/2005.

Zingiberaceae

M. Cheek (K)

Fl. Cameroun 4 (1965).

Aframomum sp. of Dom

Herb, 2m tall, gregarious, glabrous; culms densely leaves; leaves oblong, long acuminate, sub-erect; fruits ellipsoid, bright red, c. 10 × 6cm, single or in pairs, half-embedded in the soil next to the culms. Common in belt at forest transition; 1600–1607m.
Dom: Dom photographic record 35 2/2006; Cheek sight record 5 9/2006.

PTERIDOPHYTA

Fl. Cameroun 3 (1964)

LYCOPSIDA

P.J. Edwards (K)

Lycopodiaceae

Huperzia* sp. aff. *dacrydioides (Baker) Pic.Serm.
Epiphyte, slender; branches pendulous, to 35cm long; leaves
c. 10 × 2mm, narrowly lanceolate. Forest and grassland;
1633m.
Dom: Cheek 13470 9/2006.

Lycopodiella cernua (L.) Pic.Serm.
Acta Botanica Barcinonensia 31: 11 (1978); Bull. Jard. Bot.
Nat. Belg. 53: 187 (1983).
Syn. *Lycopodium cernuum* L.
Terrestrial; lower parts of stems (stolons) long-creeping, the
rest erect, much-branched; foliage leaves c. 4mm long, very
narrow; strobili very compact, c. 1cm long, drooping.
Roadsides, fallow, farmbush and forest gaps; 1780m.
Distr.: pantropical [pantropical].
IUCN: LC
Dom: Cheek 13547 9/2006.

Selaginellaceae

Selaginella thomensis Alston
Lithophyte, to 10cm, green, glabrous; stolons light brown;
frond-like portion to 10cm long, 3cm wide. On rocks in
forest; 1931m.
Distr.: Sierra Leone, Nigeria, Cameroon and São Tomé
[Guineo-Congolian].
IUCN: LC
Dom: Cheek 13589 9/2006.

FILICOPSIDA

P.J. Edwards (K)

Aspleniaceae

Asplenium aethiopicum (Burm.f.) Bech.
Terrestrial, epilithic or epiphytic fern; rhizome erect; fronds
oblong-lanceolate; stipe black or very dark brown, glossy;
lamina 1-2-pinnate pinnatisect; scales deciduous; pinnae very
acute to long caudate; sori extending along pinna-midrib
(costa) and for c. 2/3 pinna length. Forest-grassland
transition; 1600–1797m.
Distr.: pantropical [montane].
IUCN: LC
Dom: Cheek 13437 9/2006; 13662 9/2006.

Asplenium preussii Hieron.
Terrestrial, or epilithic fern, to 60cm; rhizome short; lamina
linear-lanceolate in outline, 1-pinnate pinnatisect, larger
fronds c. 40 × 12cm; pinnae ± pectinate; basal acroscopic
lobe very enlarged and almost stipitate, most lobes with a
single elongated sorus along 1/2 to 3/4 their length. Forest;
1600m.

Distr.: tropical and subtropical Africa [afromontane].
IUCN: LC
Dom: Cheek 13448 9/2006.
Local name: Nkum (Cheek 13448).

***Asplenium* sp. of Dom**
Terrestrial; rhizome erect, 40cm; funnel of 8-10 fronds,
narrowly lanceolate; stipe dark brown; lamina 1-pinnate
pinnatisect. Submontane forest; 1831m.
Dom: Cheek 13630 9/2006.

Cyatheaceae

Cyathea dregei Kunze
Syn. *Alsophila dregei* (Kunze) R.M.Tryon var. *dregei*
Acta Botanica Barcinonensia 31: 28 (1978).
Terrestrial tree-fern, to 2m or more; fronds to 2.5m long, 2
pinnate-pinnatisect; old fronds persistent (pendulous/hanging
down trunk). Forest edge; 1831m.
Distr.: widespread in tropical Africa [tropical Africa].
IUCN: LC
Dom: Dom photographic record 20 2/2006; Cheek sight
record 39 9/2006.

Cyathea manniana Hook.
Syn. *Alsophila manniana* (Hook.) R.M.Tryon
Acta Botanica Barcinonensia 31: 27 (1978).
Tree fern, to 10m, erect, slender; stipe sharply spinose; fronds
arching to horizontal, to 2.5 × 1m, 3-pinnate, old fronds not
persistent. Forest & farmbush; 1570m.
Distr.: tropical Africa [afromontane].
IUCN: LC
Dom: Dom photographic record 27 2/2006.

Dennstaedtiaceae

Pteridium aquilinum (L.) Kuhn subsp. ***aquilinum***
Terrestrial fern; thicket forming; rhizome long-creeping,
subterranean; fronds to 1.5m tall; stipe erect, the base black
(remainder brown); lamina 3-pinnate pinnatisect; sori
marginal with fimbriate indusia on both sides. Grassland and
forest; 1600m.
Distr.: cosmopolitan [montane].
IUCN: LC
Dom: Dom photographic record 21 2/2006.

Dryopteridaceae

Didymochlaena truncatula (Sw.) J.Sm.
Terrestrial, erect, tufted fern, to 2m tall; stipe and rachis with
many, broad, dark-brown scales; lamina 2-pinnate; pinnules
dimidiate, trapeziform. Forest; 1570m.
Distr.: pantropical [montane].
IUCN: LC
Dom: Dom photographic record 23 2/2006.

Dryopteris athamantica (Kunze) Kuntze
Terrestrial fern; rhizome prostrate; frond coriaceous; pinnules
obliquely cuneate at base; pinnule segments untoothed.
Forest; 1633m.
Distr.: Guinea to Cameroon, Congo (Kinshasa), Angola,
CAR and S Africa [tropical and S Africa].
IUCN: LC
Dom: Cheek 13509 9/2006.

Tectaria fernandensis (Baker) C.Chr.
Terrestrial fern; rhizome erect; fronds tufted to 1m; lamina
mostly 1-pinnate pinnatisect with 3-4 pairs of pinnae, but the
basal pair with much enlarged basal pinnules. Forest, on
rocky ground; 1600m.
Distr.: Guinea to Bioko, Cameroon, Gabon, Congo
(Kinshasa) and Rwanda [Guineo-Congolian (montane)].
IUCN: LC
Dom: Cheek 13446 9/2006; 13449 9/2006.
Local name: Nkung (Cheek 13446) & Nkun (13449).

Marattiaceae

Marattia fraxinea J.Sm. var. *fraxinea*
Very large terrestrial fern; rhizome erect, to 40 × 30cm;
fronds tufted to 4m, stiff, fleshy; stipe with brown flushing
and long white- or green-streaks; swollen base with a pair of
green to dark brown, thick, fleshy stipules; lamina ovate in
outline, 2-pinnate, to 2 × 1m. Forest; 1570m.
Distr.: palaeotropical [montane].
IUCN: LC
Dom: Dom photographic record 22 2/2006.

Oleandraceae

Nephrolepis undulata (Afzel. ex Sw.) J.Sm. var. *undulata*
Rhizome vestigial, <1cm; stolons numerous, long and wiry
(many bearing tubers); fronds tufted, erect; lamina 1-pinnate,
with 30-100 pairs of slightly crenate, elliptic-lanceolate
pinnae; bases auricled; sori semi-circular. Forest, plantations
and roadsides; 1633m.
Distr.: tropical and subtropical Africa, Madagascar,
Mascarenes, India & Thailand [palaeotropics].
IUCN: LC
Dom: Cheek sight record 16 9/2006.

Pteridaceae

Pteris togoensis Hieron.
Terrestrial, to 40cm; stipe, rachis etc. without spines;
rhizomes horizontal, creeping; frond tufted; basal pinnae
peltate; costulae with few to many spinules above. Forest-
grassland transition; 1831m.
Distr.: Guinea (Conakry) to Bioko, Cameroon, Gabon,
Congo (Brazzaville), Angola, CAR, Malawi, Sudan,
Tanzania and Kenya [Guineo-Congolian].
IUCN: LC
Dom: Cheek 13629 9/2006.
Note: according to Verdcourt in FTEA: 25 of Pteridaceae
fascicle (2002), *Pteris togoensis* = *P. catoptera* Kunze var.
catoptera (noted by Edwards, xi.2008).

INDEX TO VASCULAR PLANT CHECKLIST, FIGURES, PLATES AND RED DATA PLANTS:

Note: Accepted epithets are in roman. Synonyms are in italics and their page numbers in italics. The page or plate numbers (a number and a letter) of illustrations and/or photographs are in bold.